Laboratory Fu
Chemical Contain

by Chip Albright

A Creative Solutions Publication
Creative Solutions – Phoenix, AZ USA

Disclaimer

This publication is designed to provide accurate and authoritative information with regard to the subject matter covered. It is sold with the understanding that the contents do not render professional advice. If you require professional assistance or expert advice, you should seek the services of a licensed professional.

Any company names, logos, or images used that are trademarked or copyrighted by others are acknowledged to be the property of the trademark or copyright owner.

To request permissions, contact the publisher at
info@laboratoryfumehoodsexplained.com
Paperback: ISBN 978-1-7357110-1-0
Ebook: ISBN 978-1-7357110-4-1
First paperback edition October 2020.

Edited by Kaylene Ray
Proofreader: Rhonda Albright
Cover art by Kristen Andrews
Cover Photograph by Sunway
Photography by Lauren Albright and Chip Albright
Graphics provided by ICT^2

Printed in the USA

Creative Solutions
PO Box 71477
Phoenix, AZ 85850
USA
www.laboratoryfumehoodsexplained.com

Table of Contents

About the Author

Chip Albright

This book is a result of a 40 plus year journey into the world of laboratories. Over those years, Chip worked for a number of industry leaders in various roles. His introduction to ventilation products started in 1985 when he was transferred to the fume hood factory of Kewaunee Scientific in Lockhart, Texas. He designed his first fume hood in the late 1980s.

After Kewaunee, he joined Jamestown-JMP in Jamestown, NY. The first ASHRAE 110 was introduced in 1985, and a major revision was released in 1995. Much had changed as a result of advancements in fume hood testing, and VAV controls, giving room for innovation. Plus, with OSHA gaining jurisdiction over laboratories, many labs that had no fume hoods were now adding them. This was the perfect time to start with a blank sheet of paper and design the best hood we could.

That fume hood was known as the Isolator and included many innovative features including the chain and axle drive system for the sash and full pan steel superstructure. This product was the beginning of a long road of innovation.

In 1999-2000 Chip was Chairman of SEFA and a driving force for the rewrite of SEFA 1- Fume Hoods. Through this time, it became apparent that there was still ample room for advancing the art of fume hoods. Chip had moved to Collegedale, LLC in TN and took on the challenge to create the next generation of fume hoods. During this time the G^3 was developed. It was a high-performance fume hood with many advanced features. As these hoods were installed in large quantities, it became apparent that a great fume hood would only work as well as the laboratory ventilation system it was connected to. Chip developed a more holistic approach. The next few generations of the G3 focused on being more robust and capable of safe performance even on a marginal laboratory ventilation system. By 2010 it was becoming clear that there was a huge disconnect between those designing and producing fume hoods and the people who were engineering and installing laboratory ventilation systems.

Then there was that "aha" moment where he realized that fume hoods were not a standalone product even though that is how they were being sold. Instead they were the user interface to the mechanical ventilation system and until fume hoods became an integral part of the laboratory ventilation system, their performance was going to be questionable.

In 2008 Creative Solutions was created to explore how to best advance fume hood technology. In 2011, Chip became independent and began to explore the future of fume hoods. As a global consultant, he saw that in many parts of the world there was a total lack of understanding about how fume hoods worked and organizations were spending large amounts of money buying, installing, and operating fume hoods that were not providing an acceptable level of safety. After seeing the conditions that many users were working under, Chip became a user advocate and has spent much of the last decade educating users and laboratory managers about fume hood safety.

Lastly it became clear that because of the strong legacy of separation between fume hood manufacturers and engineers designing laboratory ventilation to achieve significant change, fume hoods were going to have to become smarter and less dependent on room conditions to perform safely. This has led to working on many innovative next generation smart fume hoods. Innovation has proven to be more accepted in countries like China and India where there is not as much resistance to change or regulation restricting innovation.

Chip has been a major player in creating standards though his work with SEFA and is a sought-after speaker on the subject of fume hood safety. His current focus is to continue to educate the market and the new generation of fume hood designers with the hope that meaningful change can take place in making laboratories truly safe for the user.

Introduction

Many laboratories have fume hoods installed. A chemical fume hood is a significant safety device that is critical to maintaining acceptable laboratory air quality and preventing user exposure to potentially harmful chemicals. Yet, the fume hood is the most misunderstood and misused safety device in the laboratory. So much mystery and myth exist regarding how a fume hood works and how to use it safely. It starts with the perception that a fume hood is a standalone device and if there is a fume hood in your laboratory and you use it, you will be safe. The reality is that a fume hood is only a small component in the mechanical ventilation system of a laboratory. It is the point of interface between the user and the laboratory ventilation system (LVS).

The fume hood does very little alone. Yes, there are good hood designs and not so good hood designs, but until connected to a properly designed and maintained laboratory ventilation system, the fume hood is just an enclosure that does nothing. The primary purpose of the fume hood is to protect users from exposure to harmful chemicals that are being used inside the hood. To do this, a fume hood must capture, contain, dilute, and exhaust those dangerous substances. None of this is possible without a ventilation system. Secondarily, the fume hood should provide some physical protection in the case of fire or explosion.

This book was written for informational purposes, it provides a background and explanation of how a fume hood works and how to use it in an effective way. The whole subject of fume hoods is full of misinformation, confusion and even controversy. Do a search for "fume hood books" and the only book you are likely to find was written by G. Thomas Saunders back in 1993. Why is there so little written about a subject as important as fume hoods? While there are standards and best practices, web posts and media articles that reference fume hoods, there is nothing from the 21st century that offers a comprehensive, factual discussion of the subject.

Having been in the industry for over 40 years I have seen it all -- the good, the bad, and the ugly. I have designed seven major fume hood product lines. I have worked for companies that have collectively built nearly 500,000 fume hoods over those years. Some of my later fume hood designs were truly world class and had great performance in the test laboratory. But it is possible to take that fume hood out of the test laboratory, install it in a working laboratory, and the same fume hood that performed so brilliantly in the test laboratory may fail to adequately protect users in a working environment. Users see the brand name and the test laboratory reports and understandably believe they are safe while using the fume hood.

In the post sales visits to laboratories, I often saw my world class fume hood not working properly. I spent a lot of time educating all stakeholders about fume hoods and how they must be integrated with the whole building to work properly. What I found was that due to the complexity of fume hood operation and the lack of understanding about how a fume hood works, no one really wanted to tackle the problem.

Through this book I hope to provide a needed source of accurate and timely information about fume hoods and their use in the laboratory environment. My goal is to cover the following points:

- the purpose of a fume hood is to capture, contain, dilute and exhaust harmful chemicals;
- fume hoods are simply a component of a laboratory ventilation system and do nothing alone;
- laboratory ventilation systems are about airflow;
- adequate testing – as manufactured (AM), as installed (AI), and as used (AU) is the only way to know if a fume hood is performing effectively;
- users are a major factor in fume hood performance and safety.

During the past ten years, I have worked mostly internationally outside the United States. Much of my time has been in India and China. It has been an eye-opening experience to see what is being done in these countries where there are few standards or regulations. Without government oversight such as we have in the U.S. with the Occupational Safety and Health Administration (OSHA), almost anything goes. One of the major problems in these countries is that many laboratories are designed and operated with no mechanical air supply systems. They use natural ventilation. In other words, they open the doors and windows when they operate the hood. It is impossible to manage air pressure and airflow balance without a mechanical supply air system. The other major issue is chemical storage. There are lots of chemicals in the open laboratory and it is common for many owners to shut the laboratory ventilation system off when users are not in the laboratory in order to save energy. The problem is that with no ventilation the chemicals in the fume hood and in the laboratory contaminate the laboratory air overnight and the users arrive back to a highly polluted laboratory the following day. The pollution is evident by the corrosion seen in these laboratories. I am quite certain that users have no idea of the hazards associated with these practices.

The last decade has been a real change for me. I went from being a supplier of fume hoods to being an independent consultant, trying to help companies make better products and to do more to protect users.

A while back, I was meeting with a laboratory owner who was preparing to build a new laboratory. The owner had been talking to several well-known suppliers about fume hoods. The plan was to use natural ventilation to save money. After some discussion, I said to him, "you do realize that these hoods will not perform safely without mechanical air supply?" His response was, "it doesn't matter if the hoods work, it only matters that we have them and that people assume they work." The indifference to user safety was shocking and convinced me to become a user advocate. It is the right thing to do.

Since 2010 I have given dozens of conference presentations and seminars on fume hood safety. I have worked with many companies helping them to improve their products and understand what it takes for a fume hood to perform effectively. I am compelled to share what I know with the hope that this information will help make a difference and that in the years to come laboratories around the world will be safer and healthier places to work.

Chapter 1

What Is A Chemical Fume Hood and Why Do We Use Them?

What is a chemical fume hood?
A fume hood is a safety device. It is classified as Personal Protective Equipment (PPE). Its primary purpose is to provide the user with protection against exposure to hazardous chemicals that they are working with and to provide some protection in the case of a fire or explosion.

Chemical Exposure is one of the invisible dangers of laboratory work. Exposure can be hazardous to your health. Although the symptoms of health damage from chemical exposure may not show up for years, the harm is occurring today. The Scientific Equipment and Furniture Association (SEFA) has published an excellent guide to exposure control devices such as fume hoods entitled, "Guide to Selection and Management of Exposure Control Devices in Laboratories." This guide is a good reference about chemical exposure and the devices and systems that can be used to prevent or minimize exposure.

Because a fume hood is a fire-resistant enclosure with a sash made from safety glass, it offers some physical protection from fire and explosions. Accidents are more common than you might think. Laboratory work can involve dangerous activities and therefore increased risk of physical harm. Training and preparedness for such events is essential to protect users.

Unless you plan on wearing a respirator while in the laboratory, you should be concerned about air quality. Breathing contaminated laboratory air can cause health problems, some of which may be permanent. Additionally, air quality not only impacts human health, it can damage equipment and compromise the outcomes of scientific work due to contamination of work samples. Maintaining acceptable indoor air quality is not only a best practice, but it makes economic sense as well.

What is a fume hood supposed to do?
The primary function of a fume hood is to prevent the laboratory users from exposure to airborne chemical hazards. To accomplish this, the fume hood uses air as its transport mechanism to:

- **Capture** the contaminants being generated or released within the hood;
- **Contain** them so they do not enter the lab;
- **Dilute** them with air to a safe concentration; and
- **Exhaust** them to the outside.

Are there different types of fume hoods?
Generally, fume hoods are designed to be used with a variety of chemical contaminants, but there are also a number of special purpose fume hoods. In a later chapter, we will look at each of these and why they are special.

- Perchloric Acid Hoods
- Radioisotope Fume Hoods
- Walk-In Fume Hoods
- Distillation Hoods

It is very important to recognize that not everything that looks like a fume hood is a fume hood. There is an industry standard definition of a fume hood. Products that don't fit that definition are often referred to as ventilation enclosures. In addition, there are other non-fume hood devices such as laminar flow hoods, chemical containment devices, ductless enclosures, ductless fume hoods, clean benches, gloveboxes, isolators, biological safety cabinets (BSC), and so on. In later chapters each of these will be discussed so that you can better understand why they are different. Using the correct product for your application is critical.

Where did my fume hood come from? Who selected my hood?
Choosing, purchasing, and installing a fume hood is a complex process. It can involve a number of people. Most of these people will not be involved in the use of the fume hood. It is uncommon for the people who will work with the hoods to have an active voice in the selection of product and even less common for them to be involved in selection or design of the overall laboratory ventilation system. Those often

involved in the fume hood acquisition process include an architect, a laboratory planner, a mechanical engineer, a construction manager, the organization's chief executive or financial manager, a purchasing manager and maybe the laboratory manager. Generally, each of these parties have differing and sometimes even contradictory objectives. Because of the complexity of the overall laboratory ventilation system (LVS), it is rare that a holistic approach is used.

How was the fume hood selected?

Unfortunately, most fume hoods are selected as if they were stand-alone devices that work independently of other systems. They are selected more based on appearance than performance. Maybe the brand is important to the purchaser, but the reality is that the best fume hood in the world does nothing until it is connected to a laboratory ventilation system. A fume hood is not a stand-alone product, it is a component in the mechanical ventilation system. If a holistic approach is not taken when designing the entire system, there is a high probability that the fume hood will not perform as expected. In fact, what was purchased to be a safety device might actually become a safety hazard.

Who designed the laboratory ventilation system?

This question is actually more important than which fume hood was selected. Why? Because the fume hood will only deliver effective performance when connected to a properly designed and maintained mechanical system. Testing has shown that 50% of the hoods not performing at safe levels are failing due to issues with the Laboratory Ventilation system. Another 25% are poorly designed hoods, and the final 25% are lacking due to poor user work practices, but without a properly designed and maintained Laboratory Ventilation System, the best fume hood in the world will not work properly.

What is Supply Air?

Laboratory safety is all about airflow. To keep the laboratory/building in balance, the supply air must mimic the exhaust. Supply air is the air being supplied to the room from the HVAC system. It usually enters the room through an opening often in the ceiling called registers. Supply air is also considered creature comfort because it provides temperature and humidity control.

Most laboratories are under slight negative pressure, which means the air coming into the room, the supply air, is of a slightly less volume than the air being exhausted out. The lower pressure causes air to flow into the lab when doors are opened keeping contamination in the lab. For a fume hood to properly operate the exhaust and supply must track each other.

Understanding the requirement of air pressure balance makes it easy to see why fume hoods do not function properly with natural ventilation, such as open windows or doors. The laboratory doesn't have to be heated or cooled, but it does have to be mechanically pressurized for the fume hood to work properly.

Problems with air supply are the single biggest factor for a fume hood's poor performance -- whether it be the volume, pressure, or the method of delivery (cross drafts). Unless air balance is addressed and managed, the fume hood is likely not providing the anticipated level of safety for the user.

Many organizations have and are spending considerable amounts of money to purchase, install and operate fume hoods. Yet a sizable number of those hoods are not delivering the levels of safety they were designed for. Thus, the majority of this money is being wasted. Only when we look at the entire system in a holistic way will be able to address the issues that make many fume hoods unsafe.

Chapter 2

What Happens When A Fume Hood Isn't Working?

A fume hood is expensive to buy, install and operate. That expense is warranted only if it provides the intended protection for the user. It is easy to assume that just because there is a fume hood installed in a laboratory that it is providing a safe environment for the user. To determine if a fume hood is effective, you must understand what makes a fume hood work.

Loss of Containment

The primary function of the fume hood is to contain the chemical fumes within the fume chamber to minimize the risk that the operator will be exposed to contaminants from loss of containment. When there is loss of containment, there is a good chance the user will be exposed, either directly while working at the hood, or indirectly as the laboratory air becomes contaminated. Exposure to dangerous chemicals can have immediate effects, or it might take years for symptoms of exposure to appear. This is unlike biological exposure where the effect is more immediate. A smell of chemicals in the laboratory is a sign of possible loss of containment.

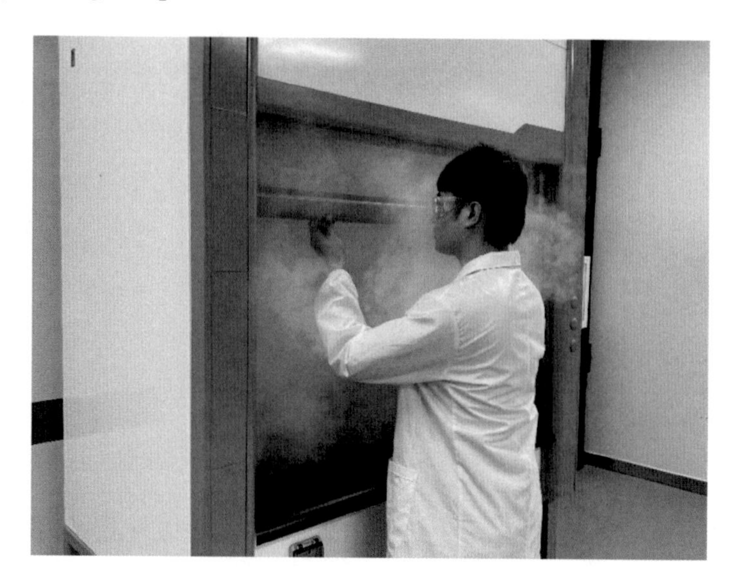

Lack of Dilution

Another failure is the lack of dilution. Lack of dilution has two components. Even hoods that are performing at an acceptable containment rate may not be effective at moving the chemicals into the exhaust stream. Many fume hoods have horizontal vortexes in the upper portion of the hood which can trap these chemicals and allow chemical concentrations to build up. This buildup can reach the Lower Explosion Limit (LEL) making the mixture explosive. If a spark is introduced to this mixture there will be an explosion or fire. In a well-designed hood, the vortex is single pass. That means that any chemicals picked up by the vortex will be exhausted after one revolution. This prevents the vortex from becoming concentrated with chemicals. There are also designs that minimize or eliminate the vortex.

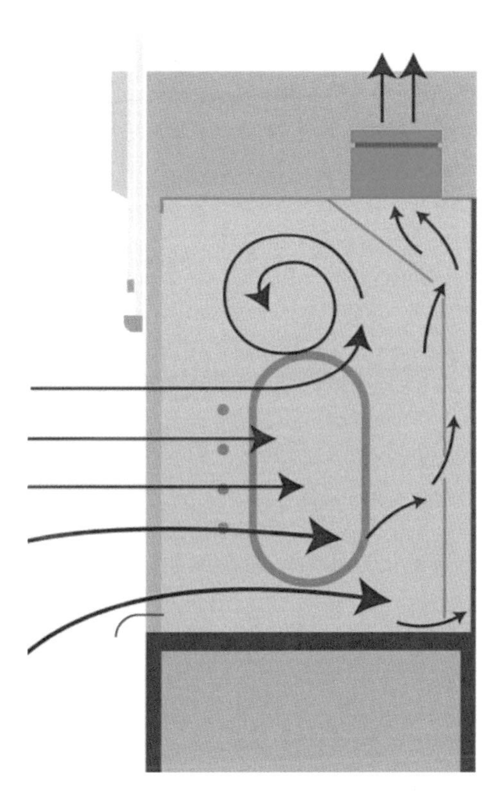

The second component is diluting the chemicals in the fume hood to a safe discharge level. The original intent of fume hoods was to capture, contain, dilute, and exhaust. In this scenario, the chemicals in the hood are mixed with air, usually room air, to dilute their concentration to a point that is safe to discharge. In the United States, the U. S. Environmental Protection Agency (EPA) sets regulations about the discharge of toxic substances. Generally, you want to dilute the chemicals in the exhaust to a safe level before discharging them. This is usually done by using air from the laboratory to dilute them. With some chemicals it is also possible to use a fume scrubber to clean the exhaust stream to a safe level. Discharging unsafe concentrations can be a hazard to maintenance personnel, to those downstream of the exhaust plume, and if there is reentrainment, to all the occupants of the building.

Laboratory Air Quality

Given the complexity of a laboratory building design, no fume hood will be able to contain 100% of the time. All hoods will suffer loss of containment at some level under different operating conditions. This loss of containment might be very slight or very infrequent or it might be very severe and constant. Regardless, the net result is the leakage of contaminants into the air in the laboratory.

In years past, we compensated for loss of containment by treating the laboratory space as a secondary containment chamber. For this to work properly, we need enough air changes to dilute the contaminants to a safe level. By keeping the laboratory under slight negative pressure and the rate of air changes high, we were able to ensure that the dilution of escaping chemicals remains at safe levels. But this strategy uses a lot of energy. As more emphasis was placed on the amount of energy used in the laboratory, we began to move away from the concept of using the laboratory for secondary containment.

To save energy, two things were commonly employed: face velocity of the fume hood was reduced and the number of air changes both in the hood and the room were reduced. While neither of these are bad strategies, it makes maintaining safe air quality a more complex and challenging task. Now to ensure safety the approach must be more risk based. What works in a low-risk situation might not be safe in a high-risk situation.

We consider a fume hood to be an exposure control device. The primary function is to minimize user exposure to potentially harmful chemicals. We are focusing mostly on inhalation hazards, breathing air that is contaminated. This contamination generally comes from three sources: loss of containment from the fume hood, chemical storage in the lab and finally from work being done outside a fume hood that should be done in a fume hood. Regardless of how the air becomes contaminated, we must look at laboratory ventilation in a more holistic way to make the lab air safe. We can safely reduce energy usage by applying ventilation strategies based on risk. This holistic approach will provide the highest level of safety while minimizing energy usage.

A poorly functioning hood is not only a waste of money and a potential health hazard, it can give users a false sense of confidence about their safety. In a later chapter we will discuss testing in more detail.

Chapter 3

Is My Fume Hood Working?

It is difficult to know at any given moment whether or not a fume hood is functioning effectively. Why is this true? Because the conditions that determine a fume hood's performance are very dynamic and everchanging. Containment depends on pressure differences. Air, like water, flows towards the low areas. Normal activity in the laboratory causes pressure shifts that impact a fume hood's performance. This can be a difficult concept to visualize, but it can be compared to watching the flames dance around in a fireplace. The flames are always moving and changing. If you have been around a wood burning fire, you know there are many factors that determine the behavior of the smoke. Loss of containment in a fireplace results in the smoke coming into the room. The purpose of the fireplace is similar to a fume hood, its function is to contain the fire and exhaust the smoke from the fire. Fire is much more visual that air. With the fire, we can see the air turbulence by watching the smoke and can observe how these factors such as the way the logs are stacked and the heat of the flame, impact the behavior of the smoke. Likewise, in the case of a fume hood, factors such as how the hood is loaded, the position of the sash, and how fast the user moves in and around the hood have an impact on performance. The difference is that you can't see it. If we could see the air interacting with the fume hood, it would be easy to see if the hood was working. Because we can't easily observe the airflow in a fume hood, we must depend on other indicators.

Remember that the fume hood is not a standalone piece of equipment. A fume hood is actually the user interface to the mechanical laboratory ventilation system. A fume hood is a single component in a very complex mechanical system. It must interact with the other mechanical systems within the laboratory and the entire building. Therefore, the fume hood's performance is dependent on how well all the other systems are operating. In Chapter 9 the laboratory ventilation system will be discussed in more detail, but the important overarching concept is that a fume hood's performance is tied to a much more complex mechanical system.

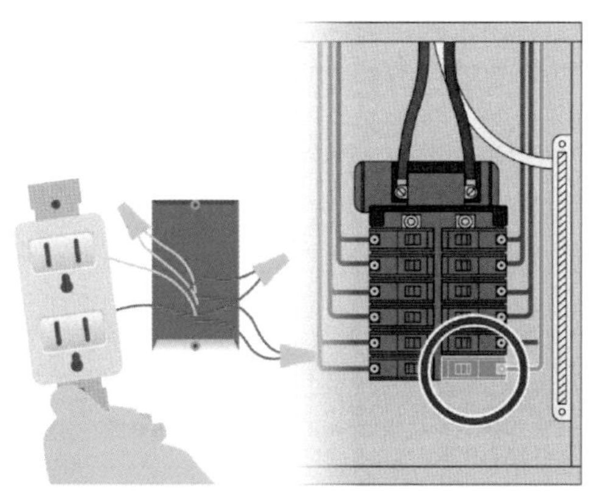

To understand the complexities of fume hood systems, comparison to more familiar systems can be helpful. An electrical system shares some of the same complexities as a ventilation system. When a device needs electricity to operate, you go to the electrical outlet and plug it in. The outlet is much like the fume hood. It is the user interface to the system. But for the power to be available, there must be wiring from the outlet to a breaker box. In the breaker box, there is a circuit breaker. The breaker box probably has wiring going to a transformer and the transformer must be connected to the power grid. Unless all of this is designed properly and functioning properly, the electrical outlet will not provide power to the device. The user's actions have an impact as well. For example, the user must ensure that the electrical device is properly rated in terms of voltage and amperage for the outlet being used. The presence of the outlet doesn't mean it is working. Depending on the circumstances, it is often necessary to test the outlet to know if is working. And if it isn't working, it is most likely not a problem with the electrical outlet, but rather a problem with some other component in the electrical system. The same is true with a fume hood, its failure to function properly is more likely a problem with the mechanical ventilation system than with the fume hood itself. And, similar to an electrical outlet, its mere presence doesn't mean it is functional or effective.

To ensure that a fume hood is functioning properly requires a holistic approach. Fume hoods must be looked at as a system. When a fume hood fails to protect users, there are many possible causes. Is it the design of the hood itself? Is it the design of the laboratory ventilation system? Is it the interface between the fume hood and the building systems such as the supply air? Is it a user error? When the system is designed all of these factors are considered. But many times, due to the complexity of the systems, what is installed doesn't function as the engineers intended. This is why testing is so important.
While there are some new technologies on the horizon, currently, the

only way to know if a fume hood is performing safely is to test it. And even then, a test is just a snapshot in time. There are a number of fume hood testing standards and there is an ongoing debate about which is the best. The various tests all give some indication of performance and some are better at evaluating certain aspects of performance than others. The circumstances under which a fume hood system is tested is important in analyzing and using the test results. Whether the fume hood was tested in a testing laboratory as it was developed, or after installation but before occupancy of the actual laboratory, or after it is being used in a working laboratory tells a different part of the story about performance. Each story is useful. A more holistic understanding of performance is useful.

By now you might be asking, why is my fume hood performing poorly? You can smell the chemicals and maybe be able to see signs of corrosion outside of the fume hood. The answer is not simple. Let's start by looking at where the laboratory is located. In North America and Europe, fume hoods generally are well designed. That's not to say they are all equal, but most have had AM (As Manufactured) testing either to ASHRAE 110 or EN 14175 standards that indicates an acceptable level of performance under perfect conditions. In these parts of the world, we would expect system problems as the cause of poor performance, but in other countries such as China and India, it is more likely that the fume hood is poorly designed and this is often coupled with a poorly designed ventilation system. In fact, in these countries, many fume hoods are installed in laboratories that have no mechanical air supply system which reduces the likelihood of effective performance to near zero. Regardless of the circumstances, the only way to know if a fume hood is performing effectively is to test it.

Chapter 4

How Design Impacts Fume Hood Performance.

Fume hood design is about the smallest of details. The difference between a good fume hood and a great fume hood is often only a few simple design features. Unfortunately, many fume hoods are designed by laboratory furniture companies rather than mechanical engineers. A major design objective by these furniture companies and lab designers is for the fume hood to coordinate with the appearance of the laboratory furniture. But a fume hood is a mechanical device not a piece of furniture!

Another problem is that many fume hoods are copies or even counterfeits of other products in the market. Even if the designer is copying a great fume hood, they usually don't have the technical knowledge to understand how the small details impact performance. As a result, they don't copy the small details correctly and what was once a highly effective fume hood becomes a mostly useless device.

To understand how fume hood design ended up in the purview of furniture companies you must look at the history of fume hoods. A more detailed history is provided in Chapter 15, but this is a brief version. The earliest fume hoods were adaptations of the fireplace and chimney. In Europe, the most common term is "fume cupboard" rather than fume hood. This is because the earliest furniture-like hoods were actually wall cupboards with glass doors placed over a window. When it was needed, the user simply opened the window. The glass doors kept most of the fumes from entering the room. Later, mechanical ventilation came along and fume hoods started to look much like they do today.

How did the modern fume hood design originate? To answer this, we first must look at the parameters that drove the designs. Height was dictated by the ceiling height of the laboratory. The hood height needed to be such that there was enough clearance for the sash to open and the ductwork to be attached. To get the fume hood into the lab, usually through a door, it was necessary to restrict the depth. The most common variable was the width. Generally, fume hoods were between

3 feet and 8 feet wide. These hood dimensions create a cubic space inside the hood. Most hoods are connected to the exhaust at the top and most work is performed in the lower half of the hood. From an airflow point of view, the space created by these dimensions is not optimal for containment.

Manufacturers often have a single model of fume hood that is available in widths of 3 feet, 4 feet, 5 feet, 6 feet and 8 feet. Even though they essentially look the same, each of these hood sizes have significantly different performance characteristics. Size matters. The fume chamber is a 3-dimensional space. The airflow within the fume hood chamber is also 3-dimentional. Changing the dimensions of the space changes the airflow. Altering the size of the chamber changes how the hood performs, yet most users would not know this. The model hood from the same manufacturer will have a different performance profile for each size. That is why it is important to have an (AM) As Manufactured test report for each size.

From the most basic point of view, containment is impacted by the airflow. Airflow patterns change with size, so if you change the size of the fume hood you change its performance. It is not a case of simply making the fume hood bigger or smaller, other details must be altered when the fume hood size is changed to minimize negative effects on performance.

How is airflow optimized? Observe how the air enters the hood, how it exits the hood, and how the airflow reacts at the various sash positions. The objective is to minimize and control turbulence and create an effective pathway for the air to capture the contaminants and then move them into the exhaust stream. The goal is to establish a smooth and mostly laminar airflow with minimum turbulence. Ideally, the fume hood design should eliminate the small eddies and the reverse flow and reduce or minimize vortexes both horizontal and vertical.

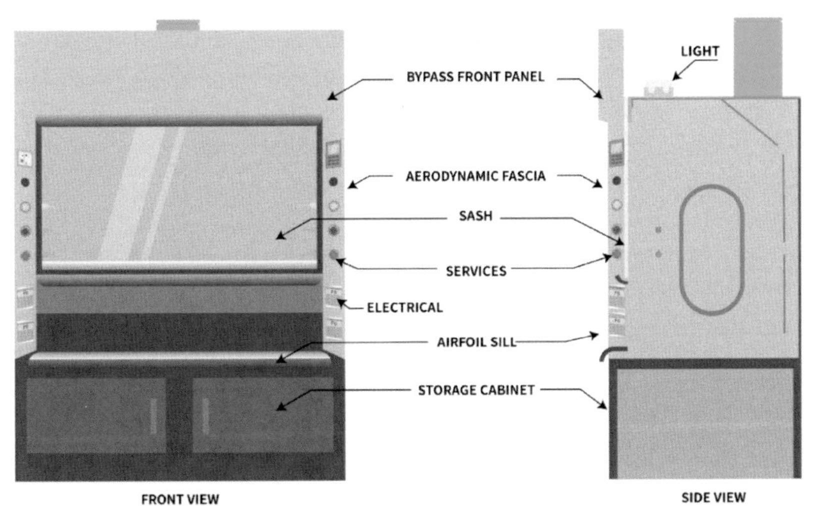

Observe how the air is entering the hood. All hoods should have a lower airfoil. The lower airfoil helps accelerate and curve the air at the bottom of the sash opening to sweep the worksurface and push contaminants to the back of the fume chamber to be exhausted. The upper airfoil has less function because it simply directs the air down the outside of the sash. When an upper airfoil is coupled with a well-designed sash handle, it directs a sheet of air inward towards the back of the fume hood. While these elements address the horizontal surfaces of the sash opening, the vertical components must also be considered. The vertical posts have an impact on how the air enters the hood. Curved surfaces are always preferred over flat ones.

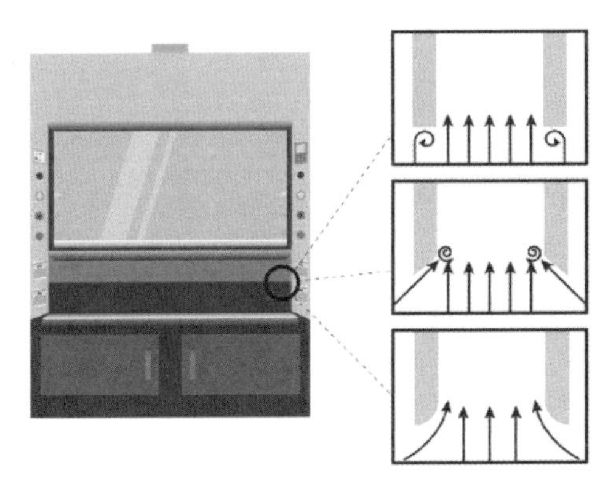

This shows the difference between a flat post, an angled post and a curved post. More attention is often to mounting the services than to the airflow.

Very subtle details can have a significant impact on how well a fume hood will perform. For a fume hood to perform acceptably, it requires a baffle system. Baffles are structures inside at the back of the fume hood chamber that create a pressurized plenum which pulls the air in because of the low pressure behind it. Because the location of the exhaust outlet in most fume hoods is at the top rear of the hood, a space or plenum is created to allow the hood to distribute airflow evenly from top to bottom and left to right. Baffle design affects a fume hood's performance more than any other design feature.

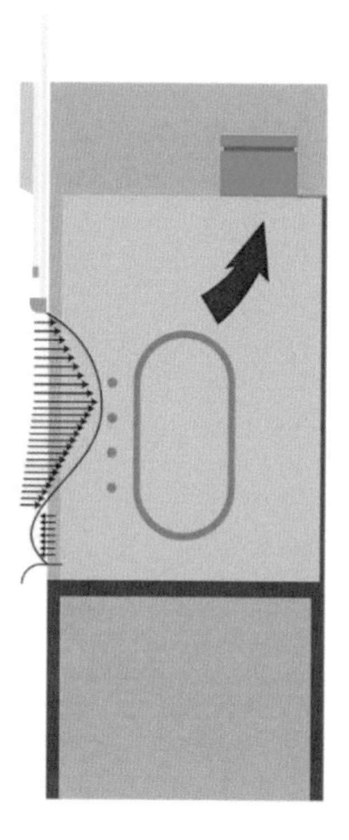

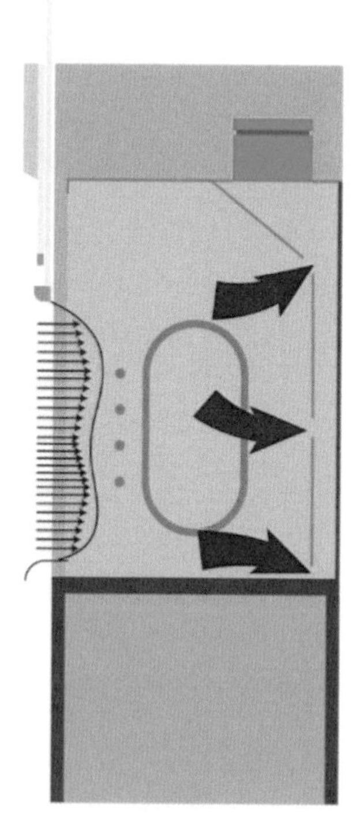

Without baffles, air just enters the fume chamber and goes directly to the exhaust duct. There is also reverse flow down at the worksurface. There are many baffle designs on the market. Some of these designs work very well and others simply don't work at all. Technical mistakes by hood designers usually involve a lack of understanding of the art and science of baffle design. Many fume hoods could achieve a dramatic improvement in performance by replacing the baffles. If a fume hood is not performing well and replacing it is not an option, upgrading the baffles might improve performance.

Because fume hood performance involves a number of design features, performance must be evaluated in a holistic way. While each feature may have a small impact, performance must be viewed in the totality. To develop a world class fume hood, the manufacturer needs the latest design tools such as CFD (Computational Fluid Dynamics) modeling and even more importantly, needs a world class test room with the latest instrumentation to validate the design.

As a buyer, it is essential to recognize that hood design is an art and a science and that some hoods are simply better than others. When considering a fume hood purchase, the more you know about how it works the better prepared you will be to make a good buying decision. While many fume hoods have good performance in a test room under ideal conditions, the same design may not perform as well under less-than-ideal conditions of the real world. What you want is a fume hood that is more robust -- a fume hood that will deliver acceptable results under a wide range of common operating conditions.

Outside of baffle design, the second most critical element is supply air. If a laboratory does not have mechanically supplied air, the fume hood cannot function properly. Natural ventilation is not appropriate for a laboratory with a fume hood. The variations in natural ventilation make it impossible for the fume hood to function properly. However, having mechanical ventilation for supply air does not automatically ensure that the fume hood will function properly. How that air is directed through the laboratory is critical. For example, the placement and type of outlets for supply air is important to minimize cross drafts that affect air currents.

The location of fume hoods when planning laboratory layout is often treated more like furniture than a mechanical system. Fume hoods are very dynamic and where they are placed makes a difference in their performance. Two of the most common placement sites for fume hoods are in corners and side by side. Such placement makes directing the airflow in the laboratory more difficult

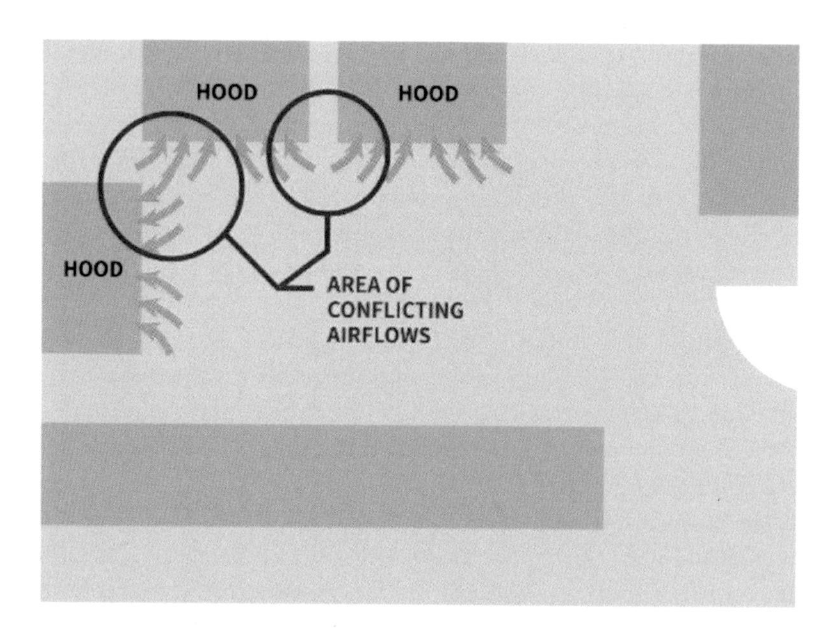

Because a fume hood's performance is so dependent on supply air and how the air is entering the fume hood, any air disruption can compromise a hood's performance. For example, locating a hood in an area that is a primary walkway or is near a door is not optimal.

Foot traffic and door swings contribute to disrupting room air patterns. The door should always open out from the laboratory for safety reasons. Should there be a fire in the hood, egress must be easily accessible. When the door opens, it allows a large volume of air to rush into the laboratory space because the air pressure in the laboratory is negative to the air pressure in the hallway. This rushing air has an effective velocity of from 2 to 5 miles per hour (175 to 450 fpm). This can not only cause cross drafts, but can also cause the room air pressure balance to change causing a shift in differential pressure and a corresponding loss of fume hood containment.

A person walking both pushes and pulls a significant volume of air. This movement disrupts the air entering the hood and can even draw air out of the hood. For this reason, fume hood locations should be away from areas of heavy foot traffic.

Hood diversity is a concept that sizes a ventilation system with less capacity than the sum of peak demand to reduce system size and energy usage. Systems are designed with a Diversity Factor, which is the theoretical maximum exhaust airflow quantity that is required at any point in time. For example, if a 50% diversity factor was used, it is assumed that no more than 50% of the fume hoods will be operating at any one time. You might ask why would we do this? The answer is simple, by incorporating a diversity factor, it enables downsizing HVAC system components and thus results in a smaller capacity ventilation system. The overall intention of applying a diversity factor when designing a VAV (Variable Air Volume) ventilation system is to achieve a lower life cycle cost. A smaller system costs less to operate.

Designing with diversity is an acceptable practice but it must be done with care. Designing for diversity differs from designing for flexibility. Systems designed for diversity may limit flexibility because they have only enough capacity to operate a predetermined percentage of laboratory chemical fume hoods at any one time. Both value engineering and sustainability can drive designs to 'right' size fans, ducting, air handlers and associated equipment.

If your laboratory was designed with a diversity factor, then it helps to know what the factor is. A hood might be working great at one moment, then start losing containment because the diversity factor has been exceeded and there is not enough air to maintain the proper airflow through the hood. When the diversity factor has been exceeded, then all operating hoods suffer, not just the last one that came online. In the case of diversity, it is necessary to manage the sash position and keep the sash closed when possible to help reduce the amount of air needed. Keeping the sash closed or opened the least possible amount reduces the amount of air that the fume hood uses. Less air also means less energy used.

Onc of the few movable parts of a hood is the sash. A good working sash is a safety feature. The sash acts as a safety shield. When working at the hood, the sash should always be closed below the face and the breathing zone. The more closed it is, the safer it is. In case of a runaway experience or a fire, the sash should be completely closed. Therefore, having a sash design that moves easily with one hand allows the sash to be closed quickly without having to stand directly in front of the hood.

There are several sash designs. Given how the air enters the fume hood, a vertical rising will provide the best performance. Use of any other design should require a very compelling need. There are a number of combination sash designs in use. These have a vertical rising frame to allow the hood to be fully open for setup. It also has horizontal panels that can be opened left to right for operation. While many people claim this sash style is safer, it isn't. When the horizontal sliding panels are opened, they create vertical vortices. These vortices grab contaminated air from the center and back of the fume hood and bring it up the sash opening making it easier to escape into the laboratory or the user's breathing zone. Sashes are discussed in more detail in Chapter 12

Performance is dependent on airflow and anything that disrupts the air patterns in the laboratory will impact performance. An empty fume hood will test differently than a loaded one. Blocking the baffles can totally disrupt the air exhaust pattern and allow copious quantities of hood-contaminated air to enter the laboratory. The setup within the fume hood will impact its performance. All equipment should be at least 25 cm behind the sash. Large objects or equipment should be elevated 50mm above the worksurface. And the baffles should not be blocked. These guidelines must be emphasized during training. But because there is no easy way to monitor compliance, the ultimate performance of the hood will depend on good user discipline.

The takeaway is that fume hoods are complex mechanical systems and a holistic approach to their design, installation and use is needed to obtain the highest performance. Unless all the components are considered, the intended safety provided by the fume hood will not be obtained. Fume hood design and location choice must be combined with the design of the laboratory ventilation system, and the physical laboratory layout in a holistic approach to obtain the highest and safest performance.

Chapter 5

What Makes A Fume Hood Work?

Fume hood performance is dependent upon the controlled flow of air in a laboratory. It is air that makes a fume hood work. Without the proper amount of air a fume hood cannot function safely. But it doesn't stop with the volume of air, it also involves the direction of the airflow and the amount of turbulence. Remember that air and water follow the same laws of physics. Air behaves like a fluid. The biggest challenge with fume hoods is that they have three-dimensional airflow. Most visualization tools, such as Computational Fluid Dynamics (CFD) modeling, only show us two dimensional slices.

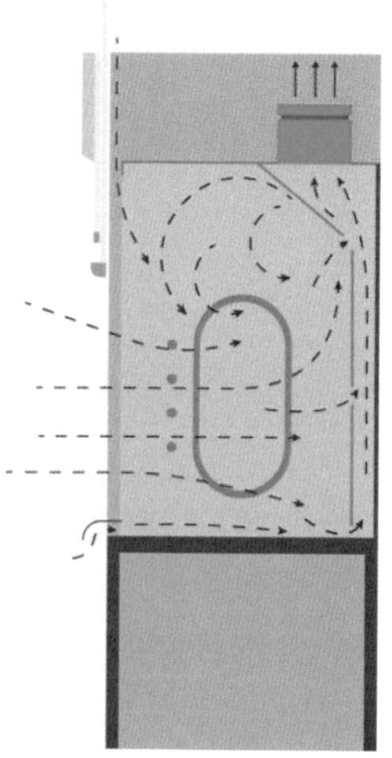

This graphic illustrates airflow. The lines represent airflow direction. If you take it at face value, then you would assume that the airflow at the bottom of the hood is laminar. But it isn't. Most CFD modeling lacks the computing power to fully model the airflow, so to get a reasonable simulation, the models decrease the precision and cancels much of the minor turbulence and eddies. This tool offers a way to validate some of the features, and tells us where to look for potential problems. But CFD modeling can only tell us so much about what the air is doing. Other testing such as smoke visualization and tracer gas containment testing are necessary to get a more complete picture of what is actually happening within the fume chamber. And the room conditions also impact the airflow within the fume hood. Much like a fireplace, there are many factors that determine where the smoke goes.

The challenges don't stop with an empty hood being tested. Most of the test protocols call for the fume hood to be tested empty. A loaded fume hood tests very differently from an empty fume hood. This can clearly be seen when doing the NIH (National Institute of Health) testing protocol which calls for boxes to be placed in fume hoods. But the testing protocol does not specify location. You can move the boxes around and see different results. Then when a person or mannequin is placed in front of a fume hood it changes everything yet again. To add to the inaccuracies, these tests are usually carried out in a near perfect test laboratory -- not in a working laboratory with less than ideal conditions.

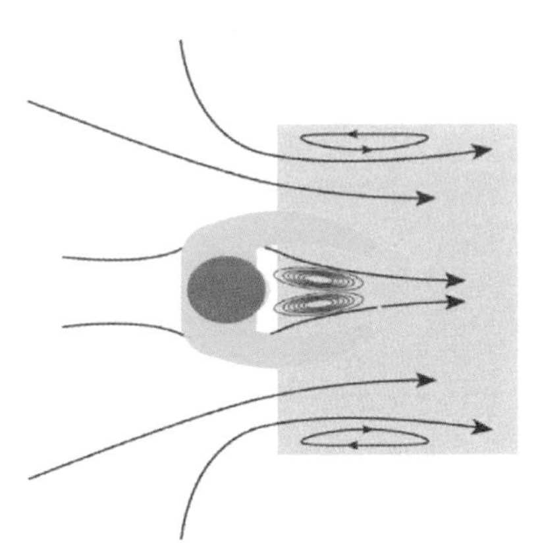

The person is much like an airplane wing and a low-pressure area is created in front of them generally causing reverse airflow.

If you are a fume hood user, have you ever worked with an empty hood or operated a hood without standing in front of it? Of course not. To be realistic, we need to look at the airflow under conditions that are more normal.

Once at a gathering of some the world's top fume hood experts, the comment was made," We don't design hoods to maximize containment, we design hoods to test well." There is a lot of truth to that statement. And if you know much about fume hood testing you know that the test protocols don't really address many of the real-world conditions. Good performance in a laboratory tested under perfect conditions, is not a good indication of a fume hood's overall effectiveness and there is no relationship to how the hood might perform in your laboratory. Fume hoods need to be tested or commissioned once they are connected to the overall laboratory ventilation system. If using the ASHRAE protocol, this would be the AI (As Installed) version of the test because a fume hood does nothing until it is connected to a ventilation system.

There are two types of pressures that are important in fume hood operations. One is static pressure. This is usually measured at the top plane of the hood. The reason static pressure is important is that it is the measurement of resistance. The higher the static pressure, the larger the exhaust system must be to create the desired face velocity. Face velocity is the speed at which the air is entering through the sash. A larger exhaust fan requires more energy to operate. So, if you have two fume hoods of the same size and you want to operate them at the same face velocity, but they have different static pressures, the fume hood with the higher static pressure will cost more to operate. The fume hood with the lower static pressure will be more cost effective. This is why the ASHRAE 110 -AM test has a very exacting method for measuring the fume hood's static pressure as well as total flow (CFM - Cubic Feet per Minute or CMH- Cubic Meters per Hour).

The more important type of pressure is differential pressure (DP). This is the pressure difference between two points. It also determines the direction of flow. Air flows from a higher-pressure area to a lower-pressure area. In the case of a fume hood, we want the airflow to be going into the hood, not coming out. We want to maintain a strong negative pressure within the fume hood compared to the room. Monitoring the differential pressure inside the hood chamber is a way to know if the hood is pulling air into the hood and maintaining containment. The stronger the negative differential pressure inside the fume chamber, the more robust the hood will be. A more robust hood will be less impacted by variating room conditions. A fume hood's design is a major factor in its ability to maintain adequate negative pressure required for containment. Hoods with less differential pressure between the hood chamber and the room are more likely to have containment issues. Most VAV (Variable Air Volume) systems add another level of complexity to maintaining an appropriate differential pressure. Today, ultralow flow face velocities are used as a technique to save energy. Ultralow face velocities also can make maintaining an appropriate differential pressure more challenging.

Recent studies have shown that air flowing over an object as small as 5-nanometers creates turbulence. A nanometer is one billionth of a meter which is much smaller than a human hair. The smallest details in a hood can cause turbulence. In most fume hoods you will find some laminar air and some turbulent air, from a performance point of view, the boundary layers between laminar and turbulent can be the most difficult to control, but controlling airflow to minimize turbulence is a critical element of good fume hood design.

If we could see the airflow it would be easy to determine if the fume hood was doing its job. We would be able to see all the air patterns and the disruptions. Seeing is believing. We could make adjustments until we had an optimal situation. Unfortunately, we can't see air, so we have to use other tools to try and understand what is happening.

There is a unique river management situation on the Ocoee River in east Tennessee (USA) that provides an illustration for understanding turbulence. In 1913 the river was dammed to provide hydroelectric power. After a lengthy legal battle, it was agreed that the middle section of the river (4.5 miles) was to be reopened for recreation daily and is home to some 20 Class III and Class IV rapids. For the 8-month rafting season, the original river is turned on and off daily. After sunset, the river is diverted to the powerhouse for use in generating electricity. At sunrise it is allowed to flow the natural riverbed so it can be used for recreation. I spent a number of enjoyable mornings watching the water rise to become rapids. When rafting the river at full flow it is hard to visualize what is under the water causing the uniqueness of each rapid, but if you watch the water rise, you get a much better understanding of how each underwater feature influences the flow of the river. Having designed many fume hoods, I often thought that if the air was as visible as the water, it would be easy to see what was really happening in the fume hood. If you could see the problems, they would be much easier to fix.

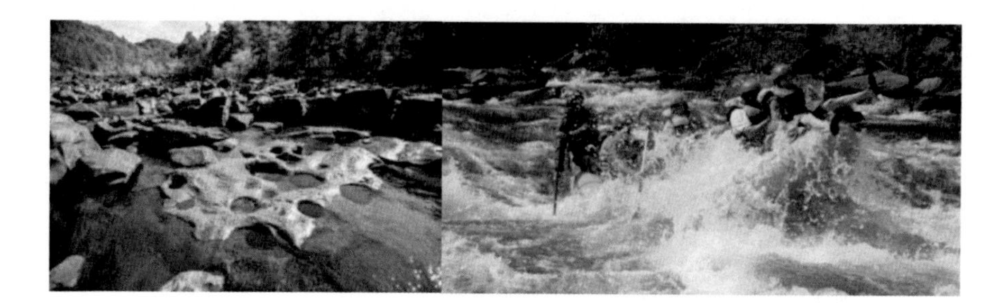

But given that air and most chemical fumes are not visible, determining what is actually happening inside a fume hood is a difficult task. While we have some visualization tools, in the end, it takes a lot of experience to really understand what you can't see. A fume hood looks simple, but it is anything but simple.

The fume hood is a dependent device. It only works as part of the overall laboratory ventilation system and is dominated by room conditions. We want the fume hood to perform under a wide range of conditions and situations, but a configuration that performs well in one situation might not do as well in a different situation. The complexities of airflow make the fume hood's job a difficult one. The more we know and understand about how a fume hood operates, the better we will be at maximizing performance and making the laboratory safer.

Chapter 6

Dispelling Common Myths Around Face Velocity

What is face velocity? ASHRAE Standard 110 defines face velocity as the average velocity of air moving perpendicular to the hood face, usually expressed in feet per minute (FPM) or meters per second (MPS). It is a simple measurement of the speed of the air going into the fume hood.

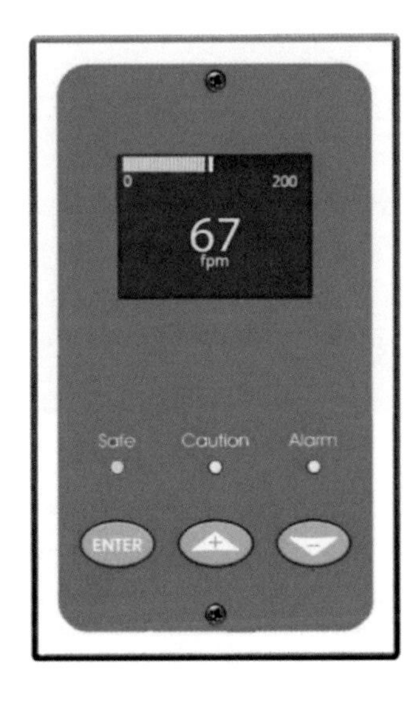

There is no relationship between face velocity and containment!

When you look at your fume hood monitor/alarm and see a face velocity reading of 0.3MPS to 0.5MPS, what is that really telling you? Many people believe that this face velocity number is an indication that the fume hood is performing safely. This is not true. The current practice of using face velocity alone as a measurement of fume hood safety is misguided. Face velocity is just one factor in the fume hood's performance and used alone it does not indicate that a fume hood is protecting the user.

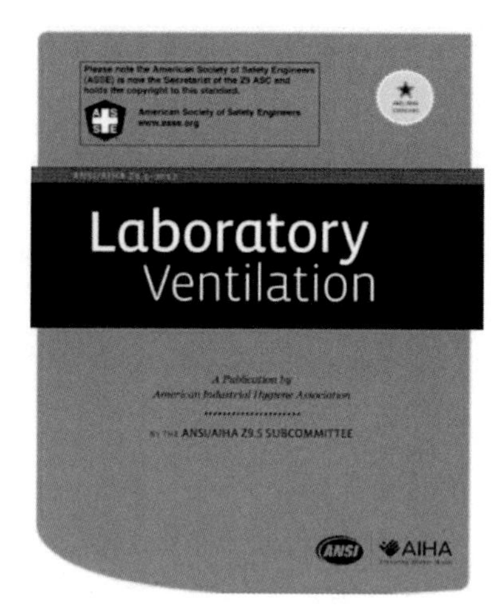

For many years, the way to make sure a fume hood was exhausting properly was to install a manometer to monitor differential pressure (DP). But because the DP is always shifting, it was hard to tell a user what the reading should be to ensure containment. Actually, observing the shifting DP numbers exposed the failure of fume hoods to maintain consistent negative differential pressure, which is necessary for containment.

In the search for a better measure of performance, average face velocity, a measurement of air speed at the inlet of the fume hood, became the standard indicator of fume hood performance. Average face velocity was more consistent than DP. It was simple to read and therefore an easier concept to sell to users. I have tested over a thousand fume hoods. Many failed the containment test. Yet those fume hoods that failed often had the prescribed range for face velocity. How can a fume hood have a 0.3MPS or 0.5MPS face velocity and still lose containment? The answer is somewhat complex. In the Laboratory Ventilation Standard, ANSI/AIHA Z9.5 on page 22 it states: "Face velocity had been used historically as the primary indicator of laboratory hood performance for several decades. However, studies involving large populations of laboratory fume hoods tested using a containment-based test like the ANSI/ASHRAE Standard 110, *Method of Testing the Performance of Laboratory Fume Hoods*, reveal that face velocity alone is an inadequate indicator of hood performance."

What does face velocity indicate? In the simplest terms, it tells us that air is going into the hood and eventually is being exhausted. It tells us that the laboratory exhaust system is functional. It is an indication of the volume of air being drawn into the hood. But air being drawn into the hood isn't an indication of what happens after the air enters the hood. What is happening in and around the hood determines the hood's performance. In testing the function of a fume hood (capture-contain-dilute-exhaust), face velocity provides little insight into whether proper containment is happening.

I was recently in a manufacturer's factory showroom. The sales person was telling me how safe their fume hoods were. So, I asked, "How do you know your hood is safe?" The answer was, "Because we have a good face velocity." This is a fallacy. Many people selling fume hoods don't really understand how a fume hood works. There is a lot of misinformation about fume hoods in the market.

Face velocity is a false indicator of safety. How did this misconception evolve? For decades, fume hoods were only turned on when they were used. Each hood had its own blower that had to be turned on by the operator. It was important for the operator to know the hood was turned on. The velocity alarm was wired to the fume hood light. When the user turned the light on, it turned the alarm on. If the exhaust fan was not turned on, a low velocity alarm would sound. Thus, having the correct face velocity was an indication that the overall ventilation system was operating. In the past, most of the fans were belt driven. It was common for the belt to stretch with use and for the fan to slow down, reducing the face velocity. Therefore, the number that could be read on the face velocity alarm was an indication of fan performance as well. However, that is not the same as fume hood performance. This disconnect between face velocity and fume hood performance has continued through the years. Just because air is moving through the hood doesn't mean the fume hood is providing safe containment.

The relationship between face velocity and containment would only be valid if the fume hood was a true laminar flow device and if the airflow was two dimensional. But airflow in a traditional fume hood is anything but laminar. Most fume hoods are very turbulent and the airflow is three dimensional.

A method of measuring fluid flow patterns (which is also applied to airflow) is known as the Reynolds number (Re). This number is used to determine whether the fluid flow is laminar or turbulent. Laminar flow occurs only at Reynolds numbers below 2000 when air flows in parallel layers, with no disruption between the layers. At low velocities the fluid (air) flows without mixing and adjacent layers slide past one another like playing cards. In laminar flow, the airflow is very orderly with all particles moving in straight lines parallel to the vessel walls. Turbulent airflow occurs at Reynolds numbers above 4000 and is dominated by inertial forces. This produces random eddies, vortices and other flow fluctuations. Face velocity tells us nothing about the turbulence within the fume hood which can lead to loss of containment.

Most fume hoods are in fact very turbulent. The reason for this turbulence is that the air entering the hood has both vertical and horizontal components. The airflow isn't two dimensional, it is three dimensional. Unlike a pipe or a duct that has a static set of conditions, a fume hood is very dynamic. Every time the sash is opened or closed the entire airflow changes. These changing airflow patterns have an impact on containment. With turbulence comes disruption of airflow and reverse airflow. This chaos offers the potential for chemicals to be pulled out of the hood and into the laboratory in what is termed loss of containment.

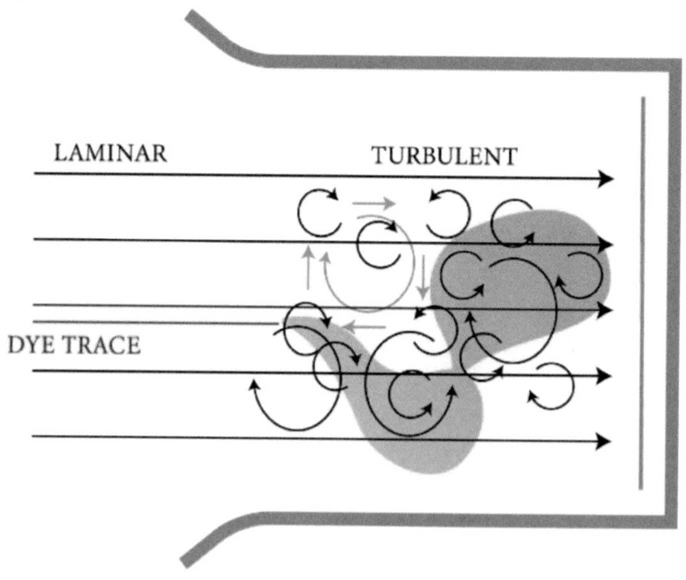

When you review testing results done by a manufacturer (As Manufactured), keep in mind that these tests were conducted under presumably perfect conditions with an empty fume hood. A fume hood that tested well in a manufacturer's test room with a face velocity of 0.3 MPS is little indication of how that hood will perform in a real-world laboratory. People buying fume hoods want to know that the devices are performing effectively and protecting them in the laboratory setting where they are being used, not in a manufacturer's test room.

Users do not work in an empty fume hood. When you load the hood with equipment and stand in front of it everything changes. To evaluate safety, you must look at much more than face velocity. It is true that adequate face velocity is necessary for the fume hood to function safely, but there are many other factors that are more directly related to containment.

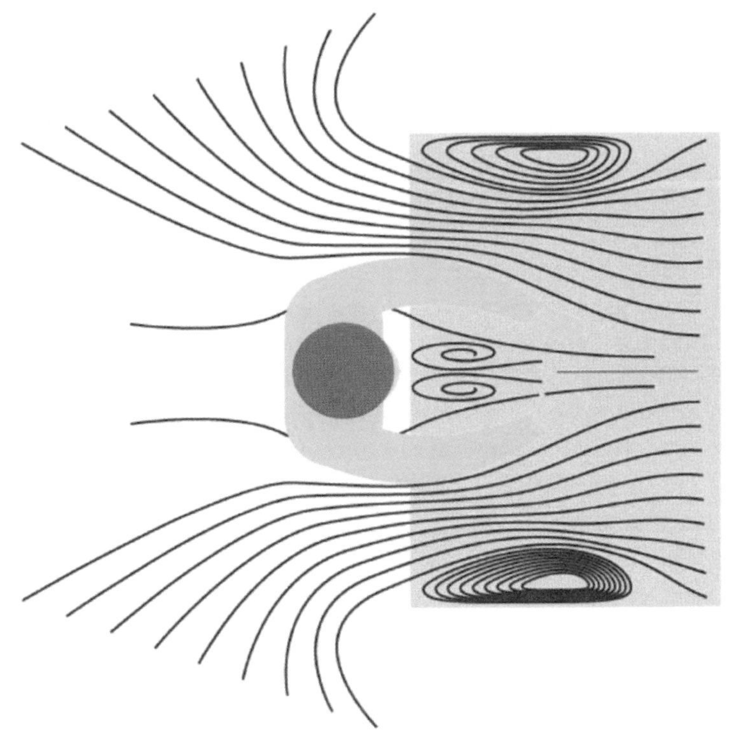

One such factor is the differential pressure (DP) between the inside of the fume hood chamber and the room. Differential pressure is a much more accurate indicator of containment. What is differential pressure? It is the pressure difference between two areas. Differential pressure determines the direction of the airflow. Air will naturally move from the high-pressure area to the low-pressure area. With a fume hood, we want the low-pressure area to be deep within the fume chamber to facilitate exhaust. But the room conditions cause the pressure to continuously shift. As the low pressure within the hood weakens, it is easier for the turbulent airflow to pull contaminants out of the hood. Monitoring the DP provides a better indication of containment than monitoring face velocity. Although DP alone is not the perfect indicator.

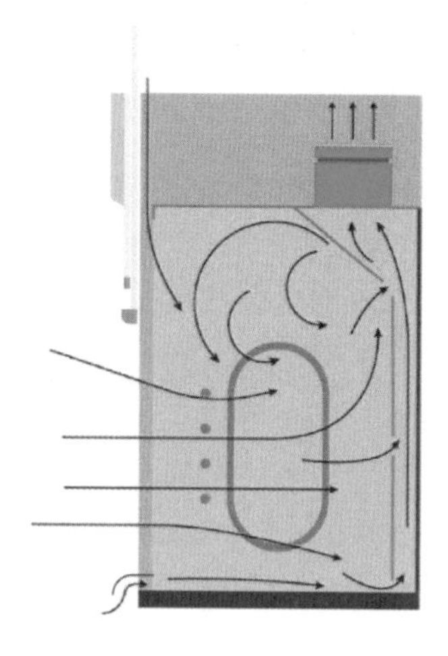

Another factor is turbulence, both inside the hood and outside the hood. Those straight lines you see in a typical CFD (Computational Fluid Dynamics) model is not really laminar as they appear in the models. These models have been greatly simplified based on model limitations and the available computing power. The airflow actually looks more like curling smoke rather than the straight lines in the model. The air is disrupted based on the mixing of the horizontal and vertical air components. All fume hoods have a degree of reverse airflow which can pull containments out of the hood during shifts of differential pressure. Measuring turbulence levels would provide an additional indication of what is actually happening to the air in the laboratory.

Fume hood tests such as ASHRAE 110-2016 AM and EN 14175-3-2019, measure fume hood effectiveness under perfect conditions. Conditions in real-world laboratories are not perfect. Additionally, in the test laboratory, only the properties of the fume hood are tested. Until the fume hood is connected to a laboratory ventilation system, there is no way to know how it will perform. Performing a test such as ASHRAE 110 – AI (As Installed) or AU (As Used) is the only way to know if a fume hood is actually achieving proper containment. Although there is technology under development that could change this.

Within the laboratory there can be a number of potential sources that can pollute the indoor air. The most likely source is a fume hood that is losing containment. Another is improper chemical storage. In addition, there is often work being done outside the fume hood on the countertops. Regardless of where the contamination comes from, the objective is to keep the laboratory air at a safe, breathable level.

This is done by replacing contaminated air with fresh air. This process is measured in terms of air changes. An air change occurs when 100% of the air in the laboratory is replaced with fresh air and is measured as ACH (air changes per hour). There are guidelines and standards that establish what an acceptable ACH is based on the type of space. These standards range from 2 to 20 ACH.

Typically, in laboratories, the recommended ACH range will be from 6 to 20. The exhaust rate and the size of the room will determine the ACH in any particular space. It is the job of the Safety Officer or Chemical Hygienist to determine what the ACH should be for a particular laboratory. If it has been determined that the desired ACH is 10, the cubic space of the laboratory is calculated and multiplied by 10 to determine how much air must be exhausted to achieve this number of air changes. Most laboratories have two components of exhaust, the fume hood and the general exhaust (GEX). To achieve the prescribed ACH the air must either go through the fume hood or the general exhaust.

Air being exhausted through the general exhaust adds little to overall safety. In prioritizing the method of exhaust, air exhausted through the fume hood(s) is preferred over general exhaust. When a fume hood operates at a lower face velocity, there may not be enough air exhausted through the fume hood to achieve the required ACH, making exhausting through the general exhaust necessary. Generally, fume hoods with lower face velocities are not as robust as the same hood with a higher face velocity. So, when choosing a fume hood, it is important to know what ACH you seek to achieve. Then, consider the size of the room, the number of fume hoods that will be operating in the room and their likely CFM. This should help you decide the best strategy. Remember, it is not the fume hood air usage alone that drives cost. It is the amount of air that must be exhausted to achieve the needed ACH that drives cost. A better fume hood that minimizes loss of containment may be a justification for lowering the ACH and accordingly, operational cost.

An energy saving strategy is to have a dual mode system. This type of system has a day mode with higher ACH and a lower ACH night mode. The practice of turning the laboratory ventilation system completely off at night and on weekends is not a good practice. An airflow of zero is not productive.

While it is very appropriate to lower the exhaust rate when the laboratory is not in use, turning the ventilation system off allows the quality of the air in the laboratory to deteriorate overnight. It may take hours after the ventilation system has been turned back on to get the contaminated air in the laboratory to a safe level. In laboratories (mostly in India) that have employed complete shut downs of the ventilation system when the laboratory is not in use, you can see signs of corrosion throughout the laboratory from the exposure of equipment to contaminated air. If the chemicals in the air are causing corrosion to painted steel and stainless steel, what are they doing to people's lungs?

Fume hoods should be the primary exhaust device in the laboratory. They are more productive in keeping the air safe than is general exhaust. ACH drives the amount of air that needs to be exhausted, but the fume hood should be the default exhaust method.

The final area impacted by face velocity is dilution. There are two factors of dilution to consider. The first is the type of contaminants that are being generated or released in the fume hood. The amount of air necessary to adequately dilute the contaminants in the fume chamber to safe levels will vary depending on what substance and how much of it is being created or used. But generally, less air and less face velocity results in less dilution.

There must be sufficient air entering the fume hood to dilute these chemicals to a safe level before discharging them outside into the environment. As the amount of air entering the fume hood is decreased, the dilution rate may be lowered to an unsafe discharge level.

The second factor of dilution to consider is the fuel to air ratio. When there isn't enough air to provide adequate dilution, it is possible for the mixture to become explosive. Even dust or powder can become explosive when the fuel-to-air mixture is right.

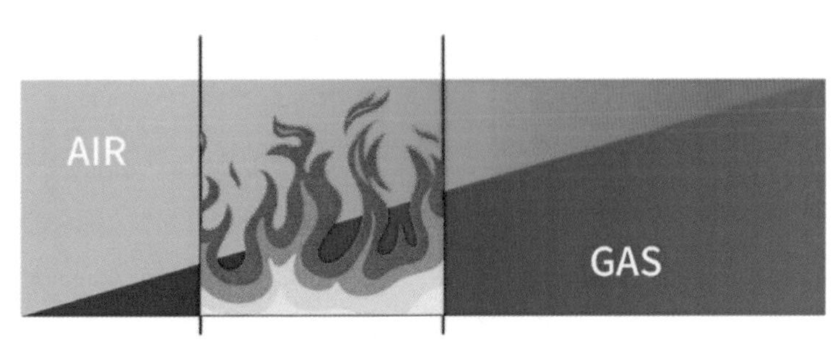

Conditions likely to cause a fire or explosion are described as the "lower explosive limit" or LEL. This is defined as the lowest concentration (by percentage) of a gas or vapor in air that is capable of producing a flash of fire in the presence of an ignition source. The LEL depends on air volume. The amount of air mixing with the chemicals determines the concentration by volume. The safest condition is for the mixture to be too lean (much more air than fuel) or below the LEL. If a lean mixture is exposed to an ignition source, nothing happens. But as the concentration builds, the mixture becomes explosive. Even as the concentration passes the LEL range and becomes too rich (more fuel but less air), it can still burn causing a fire. Less face velocity means that there is less air in the fume hood for dilution.

When looking at the LEL there is another issue that is impacted by fume hood design. Some hood designs with a strong upper vortex can develop higher fuel concentrations within the vortex. A fuel rich mixture in the vortex of a fume hood can easily cause an explosion. The design issue here is how effectively the fume hood evacuates the vortex. In some fume hood designs, the vortex just keeps turning round and round and picking up more chemicals. As the vortex picks up more chemicals that are not effectively exhausted, the mixture may cross the LEL and become explosive. While there are no formal studies, we have seen an increase in the number of hood fires and explosions in recent years as reported on the internet. This increase coincidentally correlates to the lowering of face velocities and CFM. It has therefore been surmised that lowering the face velocity has been a factor in this increase.

It bears repeating that the face velocity indicates that a certain volume of air is entering the fume hood, but it is not an indication of containment or safety. Average face velocity should only be used as one indication of fume hood performance.

Chapter 7

Loss of Containment

Loss of containment occurs when contaminants inside the fume chamber escape into the room. When a fume hood has a loss of containment, the fugitive chemicals escape into the laboratory and mix with the room air causing it to degrade. While there have been studies about the impacts of indoor air quality on human health, none have specifically focused on laboratories. Generally, these studies have focused on CO_2 levels and cleaning chemicals used in buildings. Laboratories add another level of complexity to the maintenance of indoor air quality because chemicals used in the laboratory can also become airborne.

If we use the classic definition of indoor air quality as a baseline, we can assume that the effects of fume hood loss of containment are additive to the other factors. Depending on what chemicals are in the fume hood and the lab, the impact of air quality on health varies. The second significant source of poor air quality is improper chemical storage.

Given that a fume hood is both a piece of personal protection equipment (PPE), and an exposure control device (ECD), it is clear that its primary function is to protect the user from exposure to potentially harmful substances that are inside the hood. Loss of containment is clearly a failure of the fume hood's primary function.

Also given that the fume hood is a major component in the overall mechanical ventilation system, the secondary function is managing and maintaining laboratory air quality. Historically, the fume hood was the primary containment device and the room itself was a secondary containment zone. Any loss of containment was diluted into the room air to keep the concentrations in the air at a safe level, but as ventilation technology advanced and a greater focus was placed on energy savings, this goal became more difficult to achieve.

In the 1990's, variable air volume systems were not common and the number of air changes in laboratories was quite high. It was generally accepted that the fume hood was going to lose containment from time to time and that sometimes work that should be done inside a fume hood was being done on the open bench. As a result, laboratory ventilation systems were purposely designed with a very high air change rate. These laboratory ventilation systems brought in lots of fresh air to flush the room. Although not as technically exact as modern laboratory ventilation systems, these systems probably did a very good job of providing clean air for the users to breathe, but with a high cost in energy used.

Air changes as measured in ACH are the real driver of laboratory energy costs. The more conditioned air that flows through the laboratory, the higher the energy costs. When considering safety and indoor air quality, the major factor is ACH. ACH is largely independent of the fume hood. The true energy costs of a laboratory ventilation system are associated with each air change – the cost to bring the outside air into the laboratory, to condition it, and then exhaust it. So how do fume hoods and loss of containment play into the overall costs of energy in a laboratory?

How can we justify fewer air changes and still maintain safe laboratory air quality? By looking at the risk of exposure we can evaluate an appropriate number of air changes. A better designed fume hood with less chance of loss of containment can be a justification for reducing the number of air changes. But somewhere in this search for ways to reduce the number of ACH needed, the balance between safety and energy got off center because of a failure to analyze the issues in a holistic way. As the emphasis began to shift to saving energy, we saw new concepts being put into place. Variable air volume (VAV) was one of the biggest. By controlling the amount of air a fume hood used, we could control how much additional supply air was needed to maintain air pressure balance between the fume hood and the room. To reduce fume hood air requirements, face velocities were slowed, which tends to increase loss of fume hood containment. In addition, the number of air changes was decreased, which means there is less fresh air to flush the laboratory. The air quality in many laboratories today is worse than what it was years ago.

With increased loss of containment from the fume hood and decreased number of air changes in the laboratory, safety has been compromised. The health risks of exposure to users is, of course, dependent on what substances are being used in the laboratory. All exposures are not the same. Because of the varied amount of risk from one laboratory to another, it is time to shift to a risk-based strategy. Risk-based exposure control is discussed in Chapter 13.

To design a laboratory with safe air quality, the focus should first be on determining the necessary ACH and then planning the ventilation strategy. Based on the ACH requirements, the exhaust and supply requirements can be determined, which will enable selection of appropriate ventilation products. The lower the ACH the more robust the fume hood should be. Remember, a fume hood does nothing to maintain safe air quality until it is connected to a properly designed and maintained laboratory ventilation system.

I have witnessed cases where laboratory owners installed VAV fume hoods to "save energy," only to discover that they will need to use the general exhaust to achieve the required ACH. It would have been more cost effective to increase the exhaust through the fume hood. Remember that it is the number of ACH that drives cost, not the fume hood. A well-designed fume hood with better containment can be a justification for reducing the number of air changes and thus laboratory energy costs.

Chapter 8
Re-entrainment

Where does that fume hood exhaust go? Are you breathing your own exhaust? Re-Entrainment is the situation that occurs when the air being exhausted from a building is immediately brought back into the building through the fresh air intake and other openings in the building envelope such as doors and windows.

The most common cause of re-entrainment is a poorly installed fume hood exhaust stack. The fume hood exhaust stack should be designed with the primary objective being for the fumes to be exhausted up and away from the building and other nearby buildings, giving the exhaust a chance to further dilute and disperse.

Depending on what is happening in the fume hood, the fume plume can contain highly concentrated amounts of contaminated air. There are two objectives related to exhaust. The first, is to project the plume through the exit stack at a high velocity to get it away from the building, so as not to pull the plume back into the building (re-entrainment). The second is to dilute the contaminants in the plume to safe levels by giving it the opportunity to mix with more air.

The roof can be a cluttered place based

on the design of the exhaust system.

We have all seen smoke or steam coming off of a tall stack and being blown by the wind. The same concept applies here. The standard related to exhaust in ANSI/AIAH Z9.5 requires a minimum exhaust velocity and a minimum stack height.

The exhaust stack is the exhaust tube attached to the discharge side of the exhaust blower. The goal is for the discharge velocity to be high enough to throw the discharge plume up into the air and away from the building. A building is similar to a large boulder in a river. As the water in the river hits the boulder, it creates turbulence and maybe even becomes a rapid. The wind behaves like a fluid just like the water and the building is a fixed object just like the boulder. As a result, the air around the building can be very turbulent. It is therefore important for the stack to be high enough to rise above the turbulent air.

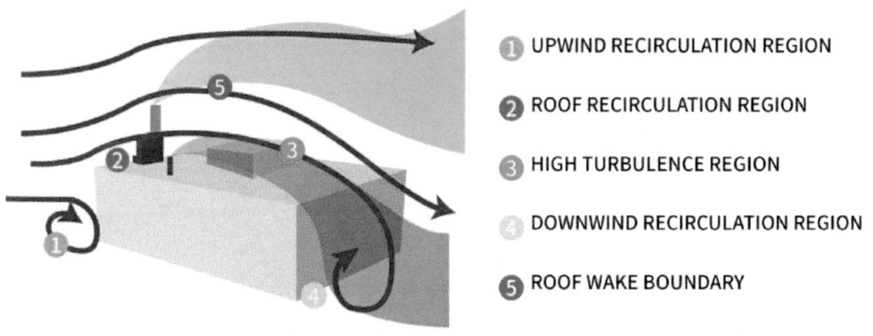

Notice in the graphic above, the spiral vortex that is created at the base of the building (4). Again, air and water follow the same physical rules. Think of this as a waterfall. At the base of a waterfall there is an undercurrent. As the water tries to rebound from the fall, it is continually pushed down by more water. This creates a churning of water. A swimmer who jumps from a waterfall can easily get caught in this vortex and be sucked down not being able to surface. The concept is similar to the chemical exhaust plume. The plume can be sucked down into the low-pressure area at the base of the building allowing it to build up concentration and be pulled back into the building through openings like doors and windows.

If the plume discharge is high enough (above 5) then the plume will join smooth air and be dispersed. If the plume gets caught in the turbulent air near the roof, then there is a good chance that it will be drawn back into the building as re-entrainment.

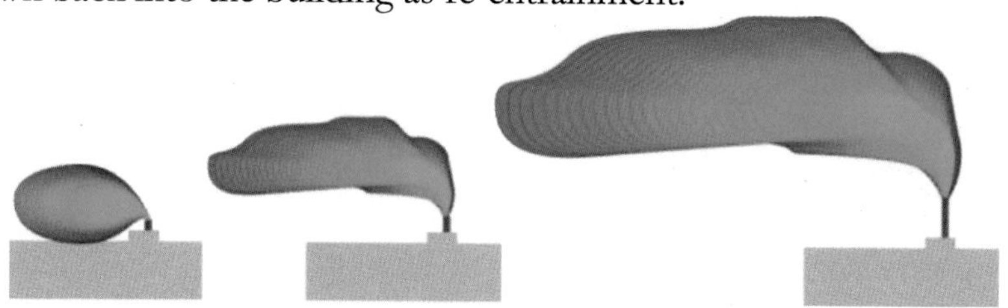

The higher the discharge is above the building the less chance there is for re-entrainment. However, the plume isn't controlled by stack height alone. It is a combination of stack height and discharge velocity.

The surrounding buildings must also be taken into consideration. The wind can blow the plume from your building into a nearby building. Or, the height of a nearby building may act as an impediment to the exhaust flow, causing it to become turbulent. An inventory of the surrounding environment is essential. The exhaust from one building can infiltrate nearby buildings.

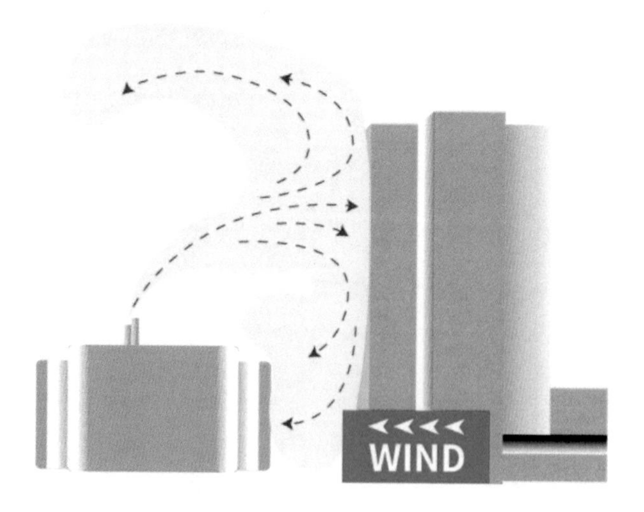

When the exhaust plume is not positioned to be taken away from nearby buildings and dispersed, it can be drawn back into those buildings through the doors, through operable windows or through that building's ventilation system.

The building roof is a busy place. In a building that contains a laboratory, not only is there laboratory exhaust on the roof, but the air intake for the supply air is also on the roof. Quite often there are also plumbing vents and vents for heaters or boilers. A lot of things are going on in the mechanical penthouse and on the roof. Again, this supports the need for a holistic approach to the entire mechanical ventilation system.

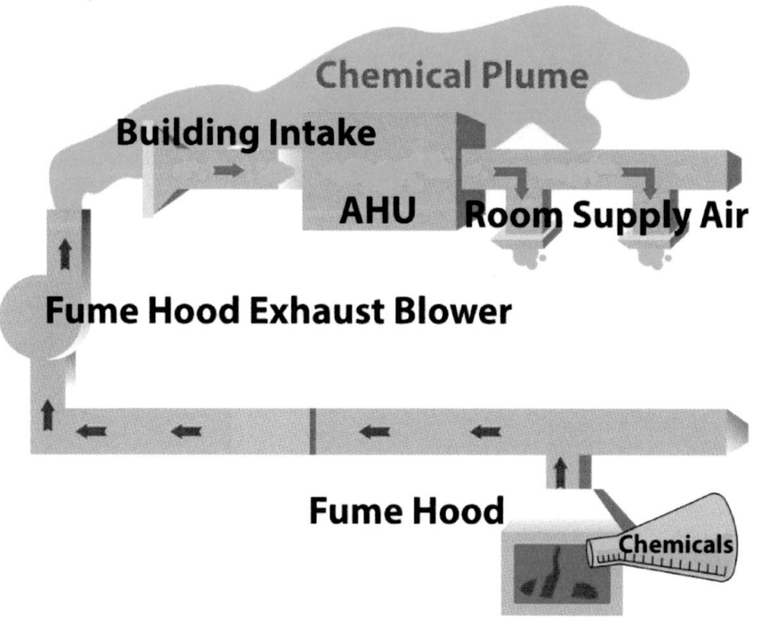

If the exhaust and supply air intake are not properly positioned and the stack height and exhaust velocity have not been optimized, the exhaust fumes can be sucked immediately back into the building through supply air intake as shown in the illustration above. This is the classic case for re-entrainment.

As a user, the most common method to detect re-entrainment is smell. When a smell of chemicals is detected in other parts of the building, you most likely have re-entrainment. Often re-entrainment is detected during routine fume hood testing. One test used to determine if re-entrainment is occurring involves releasing SF6 gas (sulfur hexafluoride) into a fume hood with the sash closed. Readings are then taken around the building, in the penthouse, on the roof, on the ground near the building and around any openings such as windows and exterior doors. This testing is very accurate and easy to perform.

Another common mistake is to place a traditional weather or raincap on the top of a fume hood stack. This defeats the purpose of using high velocity exhaust and tall stacks by redirecting the exhaust down onto the roof, almost guaranteeing re-entrainment.

To avoid this mistake, many stacks are designed with a rain water drain at the bottom of the stack where it connects to the blower. Others use a zero pressure weathercap. Whatever method is used, blocking the discharge stream should be avoided.

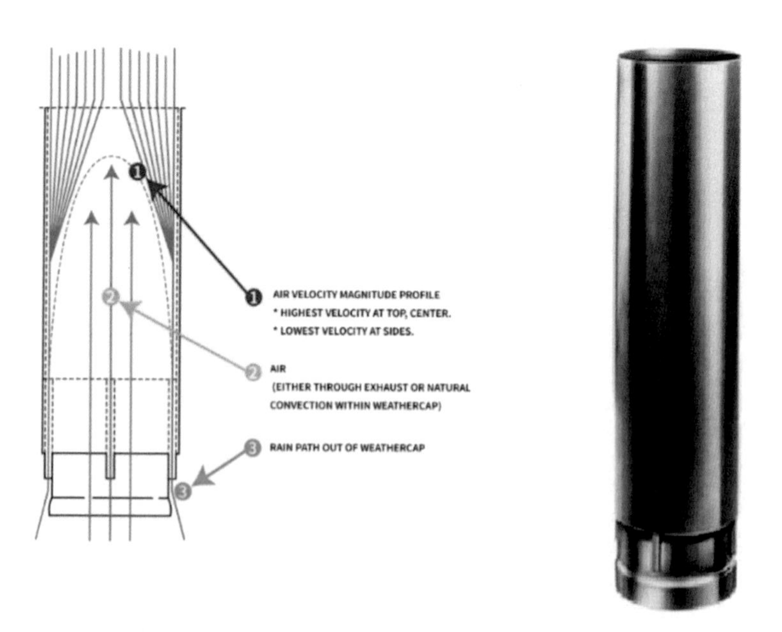

When you observe buildings and see discharge stacks on top, the fume hood stacks can be identified by their lack of weather caps. A lot of effort goes into keeping the indoor air chemical free, so don't overlook what is happening on the roof.

Chapter 9

Laboratory Ventilation System (LVS)

The best fume hood in the world does nothing until it is connected to the Laboratory Ventilation System (LVS). What is a Laboratory Ventilation System? The LVS is an assortment of mechanical devices that, at a minimum, include supply air to and exhaust air from the laboratory. LVSs can be very simple or very complex.

If you are a student taking your one and only chemistry class, then understanding the workings of the LVS is not essential, but, if you are a laboratory worker who is around fume hoods all the time, then this is a very important topic. The better a user understands the workings of the LVS the safer that user will be in the laboratory.

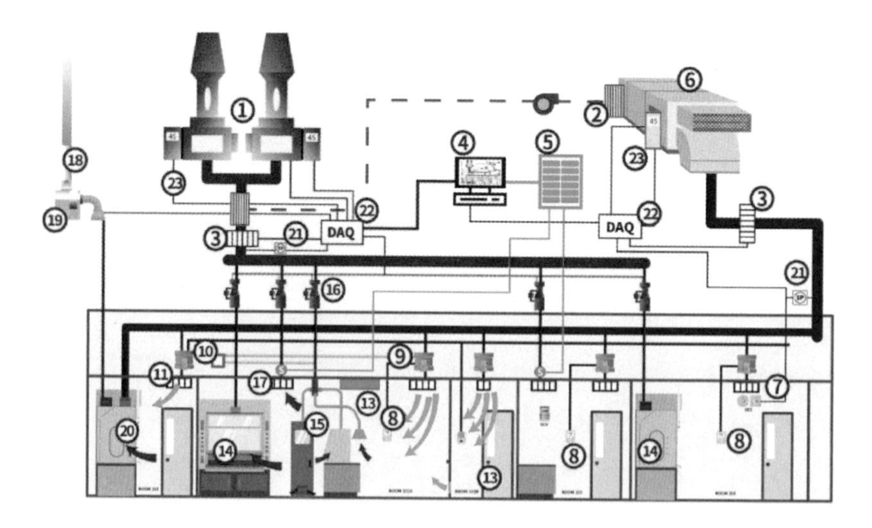

(This LVS illustration repeats on page 58 with each of the numbers defined)

The system illustrated above is moderately complex, but in the simplest of terms, the system controls the air balance in the building. When we say balance, we are talking about the air going into the building being equal to the air leaving the building. But once inside, we use the balance to control the direction of the flow of air though the building. Air will naturally flow from high pressure areas to low pressure areas.

Since labs are generally under slight negative pressure, air from the hallways usually flows into the laboratory when the door is opened. This helps direct the air from clean areas to dirty areas. The fume hood in the laboratory is at an even lower pressure than the room, so air coming into the laboratory should move towards the fume hood.

With all ventilation systems, the information essential to achieve proper air room balance is for the total flow of the exhaust and the total flow of the supply air to be equal. Unless these two flows mimic each other, the air pressure in the building/rooms will not stay in balance. Without maintaining this balance, the fume hood will not work properly. The volume of supply air must be slightly less than the volume of air exhausted to keep the laboratory under slight negative pressure.

Clean fresh air is provided by the Air Handling Unit (AHU) and is brought in by ducting to various areas of the building. In the LVS illustration, there are flow valves to regulate how much of the air is going to which areas. In a Variable Air Volume (VAV) system these valves are electrically adjustable so the amount of air can be regulated. When you are using VAV fume hoods, the exhaust varies based on sash position. Unlike a CAV (Constant Air Volume) which uses the same amount of air regardless of sash position, when the sash is open, on a VAV hood, the hood uses more air. When it is closed, a sash position sensor is used to adjust the VAV valve to use less air. It is necessary for the supply air to adjust accordingly to maintain room balance and the direction of airflow. The goal is to move air from clean areas to dirty areas. By maintaining the air in offices, hallways, and common areas at a slightly positive pressure and the laboratory at a slight negative pressure, and the fume hoods and other exhaust outlets at a very negative pressure, the air will flow from these clean areas to the dirty areas on its way out of the building.

The fume hood is on the exhaust side of the Laboratory Ventilation System (LVS), so let's first look at the supply side of the system. Whether it be a simple fan or a more complex AHU, air is collected from outside and then usually filtered and most likely dehumidified. Then it is heated or cooled. Unless the air supply in the entire building is at a constant air volume (CAV), the output of the supply system must be controlled. Today CAV systems are not very common. In most supply air systems, the volume of air is controlled using a variable speed motor. In a macro sense, the total air supply volume is monitored and kept slightly less than the total exhaust volume.

The supply air is delivered to the rooms via ducting. In designing the supply side of the LVS, the location of the supply air outlets in relation to the fume hoods is important to minimize cross drafts. Cross drafts around the fume hoods creates turbulence that disrupts the airflow into the hood. Unlike other areas of the building, the laboratory air velocity from the supply air outlets matters. In most areas of the building the high velocity outlets create no problems, but in the laboratory, high velocity outlets or registers can interfere with a fume hood's containment.

While not common in North America and Europe, many places around the world build laboratories with no mechanical supply air system. In those cases, natural ventilation is relied upon for supply air or make-up air. Fume hoods without mechanical air supply will not function reliably. Rooms with natural ventilation will have a negative impact on fume hood performance. Laboratories should not have operable windows. Unless proper and consistent room air balance is maintained, the fume hood cannot perform consistently and safely.

One of the most common problems, and a major reason for fume hoods to perform poorly, is related to the supply air. While the fume hood is an exhaust device, it will not function properly without a well-designed air supply. To maintain balance, the supply air must adjust to match the exhaust.

Keeping the exhaust and the air supply in balance is essential, but the overriding goal of the LVS is to achieve the needed number of air changes. The number of air changes per hour (ACH) is the driving factor to overall system performance and cost. Before a LVS can be properly designed, the designer must know the desired number of air changes for the particular laboratory space.

For example, a 10-foot by 30-foot room with a 10-foot ceiling contains 3,000 cubic feet of air. To get one ACH all the air in the room must be replaced in an hour. Once 3,000 cubic feet of air have been exhausted from the room and replaced with 3,000 cubic feet of fresh air, one air change has been accomplished. If 30,000 cubic feet of air is moved through the room in one hour, 10 ACH has been accomplished.

Air changes are about dilution. Any pollutants or contaminants in the air are diluted by the fresh air and exhausted. The laboratory being under slight negative pressure in effect becomes a secondary containment zone. Air enters the room either from the supply air ducts or by the opening and closing of doors. This air, because of pressure changes, is directed to the fume hood(s) or the general exhaust. Any fugitive chemicals in the air can be diluted and exhausted. In presentations, I often compare this to a prison where there are multiple levels of security to insure containment.

Equipment, people and even lights create heat in the laboratory. Air changes are also part of the process in regulating laboratory temperature. The appropriate ACH rate for a laboratory is determined by the type of work being performed (risk of exposure), chemical storage procedures, heat load, the number of people that will be in the laboratory, and the desired temperature.

In order to size the exhaust and supply units there are two pieces of information that must be gathered. They are the total cubic feet per minute (CFM) of exhaust that is required and the total static pressure loss of the system. The CFM requirements must be determined for every exhaust device in the laboratory-- fume hoods, vented chemical storage cabinets, spot extractors, canopy hoods, etc. When all of these requirements are added, the size of the exhaust can be calculated. These exhaust devices and their associated duct work create resistance to the airflow. The amount of force necessary to overcome the friction is called static pressure. To properly size the mechanical air system, both the CFM of exhaust and the static pressure requirements must be known.

The total needed CFM of exhaust can easily be converted to cubic feet per hour (CFH) which can then be used to calculate the number of air changes needed for the space. Unless the building and exhaust devices are able to maintain constant air volume (CAV), the volume of air being moved is always changing. This adds substantial complexity to the task of calculating the appropriate size of the exhaust and supply units.

Here is where it becomes even more complex. Most fume hoods are VAV hoods and the volume of air going through the hood changes based on the sash position. If there are multiple fume hoods, the total volume depends on the sash position of each hood. If the exhaust demands are changing, then the air supply must change to keep the building and laboratory air volumes in balance. Therefore, the supply of air to each room must be variable and the DP (Differential Pressure) must be continually monitored to be able to maintain the room balance. All of this has an impact on the number of ACH necessary for the laboratory. If the minimum number of air changes are not being achieved by the exhaust devices, then air needs to be exhausted through the general exhaust. Yet it is preferable to exhaust through the fume hood(s) rather than the general exhaust.

Variable air volume requires constant monitoring to maintain balance. This monitoring requires sensors, valves and other complex components. It also requires ongoing calibration. Simply installing VAV equipment won't achieve the desired cost savings and safety standards. The VAV system requires ongoing monitoring, calibrations and maintenance. A well designed and maintained VAV system can help save large amounts of energy, but without the proper commissioning, testing, calibration and maintenance, it can be a huge waste of money and may not provide the level of safety anticipated.

From the standpoint of the fume hoods, a valid alternative to a VAV fume hood is a more simplistic approach. Simple constant air volume (CAV) hoods can be used. If good, high-performance fume hoods are chosen, they can be operated at a lower face velocity (60 to 80 FMP or 0.3 to 0.4 MPS). Equipping these hoods with an automatic sash controller and occupancy sensors allows the hood to adjust the airflow as needed. There is an "in use" mode and a setback mode for "not in use" or "nights." For the LVS, a two-mode system is used which has a high and a low setting. A day mode for when the laboratory is occupied, and a night mode for when the laboratory is not occupied. In night mode, the airflow rate might be reduced to 20 feet per minute (FPM) or a reduced number of air changes within the hood. While this is a form of air volume control, it is not as complex as a fully dynamic VAV system.

Before returning to the other LVS components and issues, let's refocus on DP (differential pressure). This metric is the best indication of flow direction. The goal is to have air coming into the laboratory from the supply outlets or when the door is opened. This air is being drawn into the hood when the sash is open, and then going out the exhaust. The differential pressure is changing constantly. Opening a door, opening or closing a sash, even walking in front of a hood, can create a localized pressure shift. Good containment in the fume hood requires the hood to maintain a constant negative pressure. These continual pressure changes can cause a hood to "breathe." It is possible for the hood to inhale and exhale which causes a temporary loss of containment. As the VAV system becomes more complex, there is often a lag time in the system sensing a pressure change and adjusting to it. It may take some time for the fume hood to become stable again. It is during this lag time that there is the highest probability of loss of containment.

Twenty years ago, "low flow fume hoods," as they were called at the time, were just getting the market's attention. A large U.S. project made the decision to use these hoods. The selling point was that these low flow fume hoods used much less air than standard fume hoods therefore would be saving energy and money. When the engineering calculations were done, they were right, the low flow fume hoods used much less air. While this seemed like a good thing, it created a huge problem for this particular project. There was a government requirement for a minimum number of air changes. Had the laboratory been equipped with standard fume hoods, the hoods would have provided enough exhaust with normal use. However, the low flow hoods did not. The project was forced to use the general exhaust to achieve the required number of air changes. While the fume hoods used less air, the laboratory didn't. So, the anticipated cost savings wasn't realized. It is important to focus on air changes first before making all the other decisions.

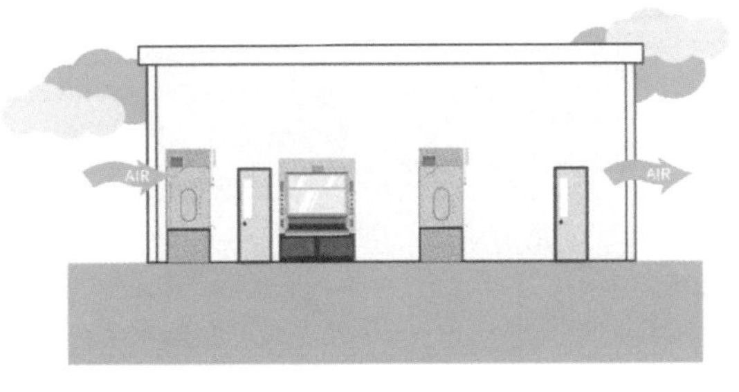

What is general exhaust (GEX)? A laboratory must have a constant airflow where the air supply and the exhaust stay in balance. When the fume hood(s) are being used, there may be enough exhaust to provide the needed air changes and keep the airflow in the room in balance. With VAV hoods, the volume exhausted is regulated by the sash opening. If all the sashes are closed the exhaust volume is very different than if all the sashes are open. Given this variance, there may not be enough air exhausted through the fume hoods to maintain room balance and achieve the required ACH. When the fume hoods are not exhausting enough air to maintain air changes and balance then air must be exhausted through the GEX. The general exhaust is a way to exhaust more air when the fume hoods are not providing enough exhaust to balance the airflow. It is more productive to exhaust through the fume hood than the general exhaust. Don't overlook the importance of the ACH requirement in the design of the Laboratory Ventilation System.

I want to comment on a couple of bad practices I have seen. The first is completely turning off the LVS when the laboratory isn't occupied. This clearly isn't a best practice. In reality, the system needs to run 24 hours a day/7days a week. Ventilation can be cut back to almost nothing during the unoccupied hours, but some air movement is needed to keep the fume hoods exhausting and chemical concentrations from building up in the labs. The result of shutdown is often seen in the form of corrosion around the laboratory. Not only can this chemical build up damage equipment, but can harm people who return to the laboratory before the contaminants have been purged. Plus, contaminated lab air can contaminate experiments.

The other bad practice is having fume hoods in laboratories with no mechanical supply or make-up air. These laboratories have only natural ventilation, such as opened windows and doors. This makes continual containment impossible. With natural ventilation there is no way to manage the supply. The air pressure will continually attempt to reach a point of equilibrium. This makes it hard for the hood to maintain negative air pressure. Also, when windows are involved, there is often excessive cross draft that impacts the fume hood's ability to maintain containment. If mechanical air supply cannot be provided, then this is a case where a Synchronized Supply hood should be considered. A Synchronized Supply Hood is a new class of fume hoods that introduce supply air directly into the fume chamber. Both the supply air and the exhaust have VAV valves to keep the DP within the hood negative to the room. This type of system synchronizes the supply air with the exhaust to maintain a stable airflow.

One of the considerations in designing the Laboratory Ventilation System is how the system will respond if the exhaust blower stops working. This can occur because of loss of electricity or a blower going offline in need of service. Many laboratories have backup power to at least keep the exhaust blowers operating. On large VAV systems, the critical sensors and controllers usually have backup power. If there is no backup power, procedures to maximize containment until power can be restored should be established. In the case of mechanical issues, redundant blowers are being used more often. Today, you see fewer systems with one fume hood and one blower. It is more common to have several fume hoods on a manifold system.

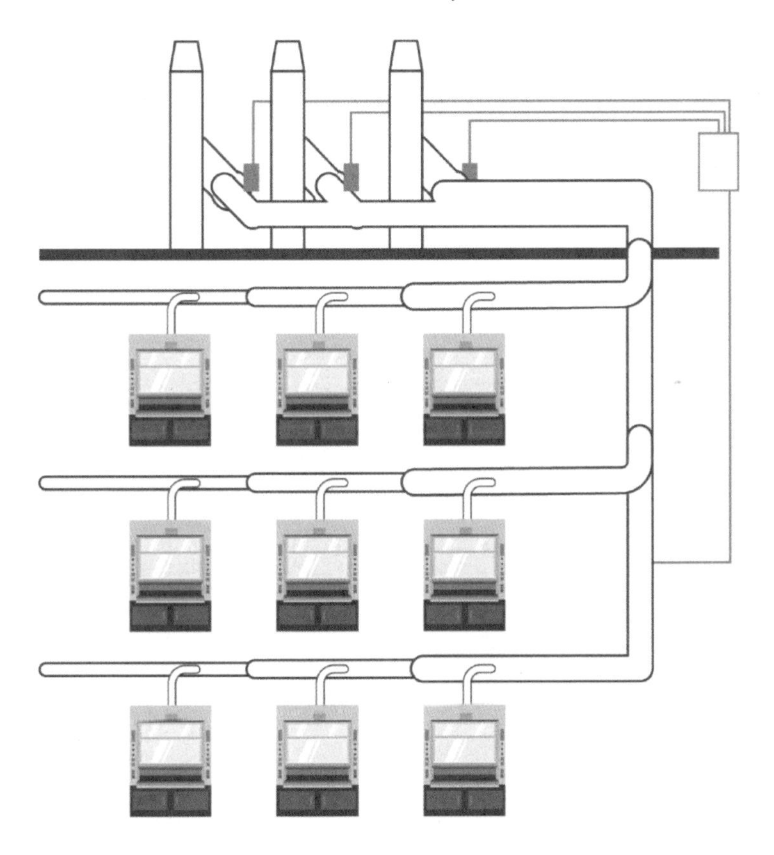

Multiple exhaust blowers or exhaust fans as they are often called, provide a number of potential advantages. First, if a blower fails or needs to be taken offline for service, this can occur without the necessity of shutting the entire system down. Even more importantly, the system can be sized so that one blower can supply much of the anticipated demand. When that blower gets close to its capacity, the second blower can engage. This also has the possibility of making the system more adaptable, so that if the overall demand goes up in the future, the second blower could run more often. This concept also works on clusters where multiple blowers are installed, but one is reserved for use during maximum demand only. Having two blowers is more cost effective from an operational point of view, because the larger the blower the more energy it takes to run it. Two or more smaller blowers that only run when needed, can save energy and provide redundancy.

Another concept used in designing Laboratory Ventilation System is called diversity. Rather than designing a system that is large enough to support all fume hoods with their maximum exhaust requirements, diversity is a concept that assumes that 100% of the possible demand will never be needed. This is because not all fume hoods will be in operation at the same time. Some will be in use with the sash open, some in use with the sash closed, and some not in use at all. If we assume that no more than 50% of the fume hoods will be operating at one time, we can design the hood with a 50% diversity factor. That allows for a smaller system that uses less energy. This is a widely used concept, and it generally works well, you need to know if this concept can safely be applied to your laboratory. If your laboratory was designed for 50% diversity, but 80% of the fume hood capacity is being used, there simply isn't enough air to go around. Some or all of the fume hoods will slow down, which increases the likelihood of loss of containment. It is important to know if your laboratory ventilation system was designed using diversity. If so, you should know what the diversity factor is. This information can help you establish appropriate safety procedures.

There are two main types of blowers or fans, that are generally used. On smaller systems, a centrifugal or axial fan will likely be used.

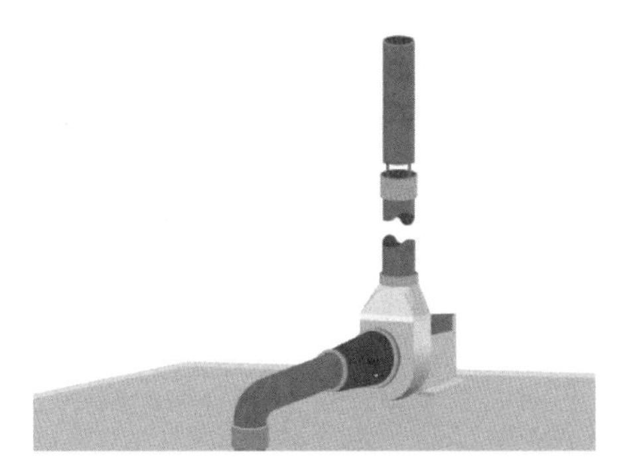

With this style of blower, the bigger the motor the more air is moved. This type of fan usually has a variable frequency drive (VFD) which controls the speed of the motor. When less air is needed, the motor can be slowed down. When more air is needed, the speed of the motor can be increased.

The other style of blower or fan is the high plume dilution exhaust fan.

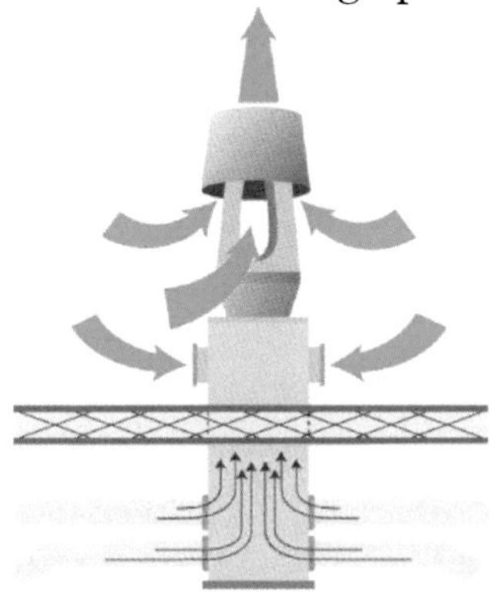

This style of blower not only pulls the exhaust air out of the building, but it also mixes it with outside air to further dilute it. These blowers have a very high discharge velocity because outside air is being added. This allows for a high exhaust plume without the need for a high stack.

When you are working at your hood, can you hear the fume hood running? What do you hear? Is it motor noise you hear or is it the sound of air rushing into the exhaust outlet? Whatever noise you are hearing it is most likely the ventilation system and not the fume hood. Fume hoods themselves do not generate noise; fume hoods can amplify the noises that are generated by the exhaust system.

When we test for noise in a test laboratory, a sound meter is used to take a reading just below the exhaust duct without the hood connected, then the hood is connected and another reading is taken. The sound reading should not increase by more than 10db with the fume hood connected. Without the fume hood connected, most of the sound is created by high air velocity and turbulence within the duct. A noisy fume hood is actually caused by the ductwork rather than the fume hood. Air speed in the duct has a big impact on the noise level. The higher the duct velocity, the more noise, but an optimal duct velocity must be maintained. If it is too fast, it creates turbulence and noise and increases static pressure. If it is too slow, the contaminants in the airstream can fall out of the airflow. The optimal air speed in the ductwork is between 1300 FPM and 1600 FPM.

In some situations, dilution, through adding room air, may not be sufficient to adequately reduce chemical concentrations in the exhaust stream. There are two alternate concepts that can be used to further clean the exhaust stream. One is to use a scrubber, usually water based, which causes the chemicals to attach to the water, removing them from the discharged air. To a degree this just moves the pollution from air pollution to water pollution that may require waste water treatment. The other method is to filter the air with either HEPA or carbon-based filters. This actually removes the contaminants from the exhaust, but then the filters become hazardous waste and need to be incinerated. There is no perfect answer, other than to use less of the hazardous chemicals.

Now we are ready to dive deeper into some of the terms and concepts you might encounter. Refer to the LVS graphic on the following page (a repeat from the beginning of the chapter). Not every system is designed this way, but it will give you a general idea of how the components of a laboratory ventilation system work together.

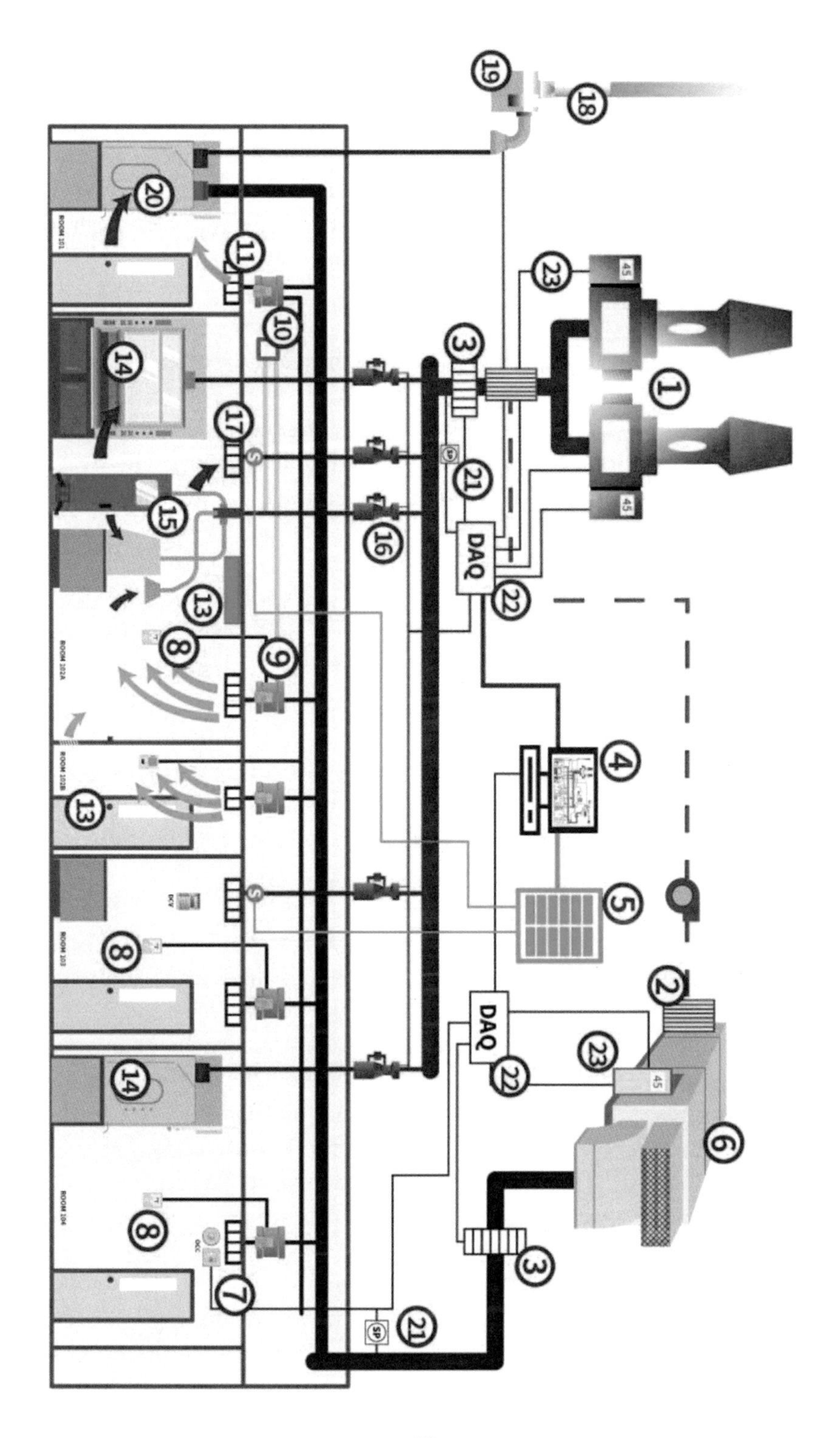

1
2
3
4
5
6
7
8
9
10
11
13
14
15
16
17
18
19
20
21
22
23
DAQ
DAQ
45
SP

Dual exhaust fans (1) – there are two things here to note, both the dual blowers and the type of blower. These are induced flow, venture type, exhaust fans and stacks with continuous power back up. Redundant venturi fans enable better plume discharge at lower flow and better system dependability. The combination of the variable frequency drive (VFD) and outside air bypass dampers (OABD) optimizes the ability to reduce flow, but will maintain sufficient plume discharge.

Energy Recovery System (2) – Various types of energy recovery systems can be used depending on the hazards in the exhaust.

Flow Monitors and System Sensors (3) – Exhaust and air supply systems often use flow stations and static pressure sensors. The measurement of total flow in addition to system static pressure improves system flow tracking.

Building Automation System (BAS) (4) – The LVS is actually part of the entire building ventilation system. The building automation system can monitor operation, detect operational problems and trend operation over time to provide useful operating metrics.

Contaminant Sensing Demand Ventilation Controller (5) – This is a concept that samples laboratory air and changes the ACH based on the quality of the air. This is the main controller and it interfaces with detectors in each laboratory.

Air handling Unit (AHU) (6) – Air handling units provide 100% outside air supplied to the laboratories that handle airborne hazards.

Occupancy Sensors (7) – The occupancy sensors detect the presence of people in the laboratory and enable flow to be reduced or setback when laboratories are vacant. This flow reduction lowers the air change rates. This lower rate of ACH potentially allows for the accumulation of concentrations of contaminants in the room which could exposure users upon re-entering the laboratory prior to purging.

Temperature Sensor (8) – The temperature sensors (i.e. thermostats) detect room temperature and translate a signal to the air supply discharge temperature controller for maintaining room temperature specifications.

Supply Air Valve (9) – The supply air for each room has a VAV valve. When the pressure sensors detect a pressure change, the room controller sends a signal that operates the valve to allow the correct volume of air to maintain room balance.

Room Controller (10) – The room controller monitors a number of factors within the room, but its primary function is to control the supply air to keep the room properly balanced.

Supply Diffuser (11) – The supply diffuser is the opening that allows the supply air into the room. In the laboratory it is preferable to avoid the high velocity diffusers because they can cause cross drafts in front of hoods. There are a number of types and materials available for laboratory applications.

Anteroom or Vestibule (12) – This is a room that serves as a secondary isolation to prevent migration or escape of airborne hazards to non-laboratory spaces.

Airborne Contaminant Filtration System (13) – These are ceiling mounted air cleaners. They contain carbon filters and they remove many chemical contaminants present in the laboratory air. The air is drawn into the unit and cleaned and then returned to the room. It has a net zero effect on DP or room air volume balance.

Variable Air Volume (VAV) Fume Hood (14) – This exposure control device is a VAV fume hood with a VAV valve and controller.

Local Exhaust Ventilation (15) – This group of exposure control devices includes vented chemical storage cabinets, gas cylinder cabinets, snorkels or spot extractors.

VAV Exhaust Valves (16) – These devices are comprised of sensors, actuators and flow dampers to modulate exhaust flow to satisfy the flow requirements of the VAV fume hood or other exhaust devices in the laboratory.

General Exhaust(GEX) (17) – This is a room exhaust used to help balance the room and to achieve the desired ACH. When the fume hoods are not exhausting enough room air, the GEX exhausts to maintain balance at the specified air changes.

Exhaust Stack (18) – This is the exhaust from the fume hood blower. It should be a minimum of 10 feet above the roof.

Fume Hood fan/blower (19) – This blower is connected to only one hood and is controlled by a switch on the hood.

Synchronized Supply Fume hood (20) – This is an exposure control device also known as a chemical containment device (CCD). It has both exhaust and supply air both with VAV valves and a controller to manage both and maintain the correct differential pressure within the hood.

Static Pressure Sensor (21) – This sensor is measuring the pressure loss in the ducting system. This is used to help determine airflow.

Data Logger (DAQ) (22) – This device is taking data from the various sensors and often analog and converting to a digital format that a computer can understand. This is often the interface between the sensors and the Building Automation System (BAS)

Variable Frequency Drive (VFD) (23) – This type of motor and controller allow the motor to operate at various speeds. When the demand is low the motor can slow down and as demand increases the speed of the motor which correlates to volume of air that is being moved.

Chapter 10

The User Is A Critical Part of The System

What happens or doesn't happen in a fume hood can make it unsafe. You would expect people that work in laboratories, because of the very nature of their education and training, to be highly disciplined when it comes to all aspects of their laboratory operations. However, that is often not the case. I have observed fume hood users doing things that make it nearly impossible for a fume hood to work properly. In most cases, these laboratory workers simply don't know any better; they have not been properly trained on how a fume hood works and how users should interface with it to maintain safe operation.

Recent news articles quote studies claiming that as many as 45% of laboratory users say they received no training on laboratory safety, including the proper use of a fume hood. Like all things safety related, the user is a critical part of the system and the outcome. For instance, safety glasses don't work if the user doesn't wear them.

Fume hoods are aerodynamic devices. Their performance is influenced by airflow and turbulence. What happens in and around the hood impacts containment. Therefore, good operating procedures greatly increase the probability of containment and reduce the risk of exposure to the user and other laboratory occupants.

If there are air quality problems within a laboratory, you may not be able to change the fume hood or upgrade the laboratory ventilation system, but you can train users to maximize the equipment they have by adopting good work practices.

When fume hoods are tested AU (As Used), many hoods fail to perform at safe levels. About 25% of those failures occur because of user practices. The corrective action is to train the users and help them understand how to do their work in a way that maximizes the fume hood's ability to contain hazardous materials.
Many users believe that a fume hood is protecting them simply because air is moving through it. They have no reason to question the performance. They do not realize that activity in and around the fume

hood impacts its performance. Fume hood users need training and knowledge to be an active participant in safe fume hood operation. After all, it is their health and welfare that is at stake.

As I say in my presentations at industry conferences, safety is, "no accident." Working safely with a fume hood requires a plan--knowledge of what you are doing and what could go wrong. For example, if a runaway experiment causes a fire inside the hood, do users know what to do? Are there standard operating procedures (SOPs) for this event? Do you close the sash and get the closest fire extinguisher? Do you call 911 (emergency services)? Do you alert your supervisor? There are many possible actions you could take and many different outcomes. Some of the outcomes are not good. To ensure the results you want, there needs to be a plan and the user needs to know, understand, and follow the plan.

The actions of fume hood users affect everyone in the laboratory. Air quality affects everyone. If there is an accident, it puts everyone at risk. It isn't just about your health and safety, when fume hoods fail to contain and contaminate laboratory air, that puts everyone in the lab at risk of exposure. The poor air quality can damage equipment and compromise the results from work being performed.

I have long supported user training and certification before a user is allowed to use a fume hood. While some laboratories have adopted training and certification programs, it is not currently a requirement. As fume hood technology becomes more advanced, it will be necessary for users to receive extensive training before the hood will operate for them. There are fume hoods in use that will not operate unless the operator has the proper credentials in the form of a RFID ID or biometric identification, for now, simple training is a step in the right direction. If we are not planning to succeed, we are likely to fail. Lack of proper planning and poor procedures is the major cause of laboratory accidents.

The following questions are just a few that a user should consider to achieve a safe fume hood experience:

- Is the hood face velocity adequate?

- Are the back-baffle slots clear of obstruction?

- Is the work positioned at least 6 inches into the hood fume chamber?

- Is my hood housekeeping good?

- Is the bottom front air foil in place?

- Does my sash operate smoothly?

- Am I keeping the vertical sash closed to a point below my shoulders when I use the hood?

- Am I keeping my horizontal sash partially closed so as to obtain the smallest possible, but reasonable, opening?

- If my fume hood is indicating a low-flow alarm, do I notify my supervision?

- Can I rely on the noise I hear when standing at the hood face to tell me that the hood is working properly?

- Do I know what to do if I have a fire in the hood?

- Have my health and safety personnel performed a qualitative or quantitative evaluation of my hood, and if so, when and what were the results?

- Do I smell chemicals in the laboratory when I first come in?

- Do I see rust/corrosion around the laboratory or in the hood?

- Do I know the risks or potential issues with the experiments or work I am doing?

- Have I been trained to properly operate the fume hood and did I understand the training?

- Does the laboratory have safety procedures? Do I understand them?

- If there is an accident, do I know what to do?

Years ago, I started a "train the trainer" program and it has produced good results. I try to create a high level of fume hood expertise in the companies I work with. Training is only effective when the trainers know and can disseminate accurate information. I can't say enough about how important it is to educate and train users so they can be an active and functional part of the system. Once education and training occur, management's job is to ensure compliance. Where training has occurred, the biggest failure comes from the lack of user discipline.

As a user, you are a key component in the system that is there to protect you. You can make a difference in laboratory safety. It is your health and safety that are at risk.

Chapter 11

Fume Hood Testing

The only way to know if a fume hood is performing effectively is to test it. While there is debate about which test is best, any testing is better than no testing at all. This chapter contains an overview of some of the more common tests and explains why testing is so important.

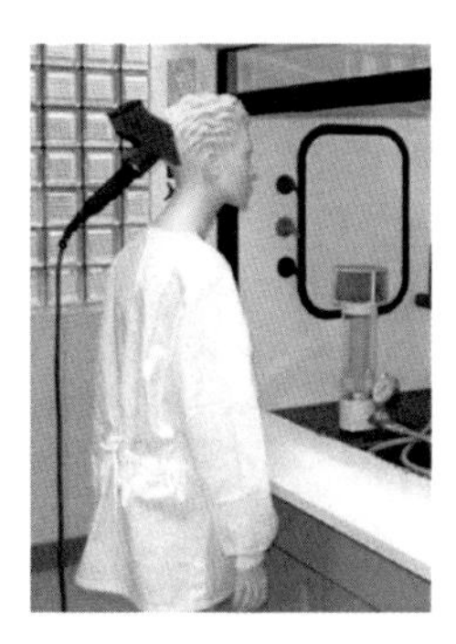

It bears repeating that a fume hood does nothing until it is connected to a laboratory ventilation system (LVS). The only way to know if a fume hood is working properly is to test it after it has been installed in your lab. A fume hood alone can't be tested because it is not a standalone device. It requires a mechanical ventilation system to function.

The laboratory mechanical design involves engineering not only the laboratory space, but also the entire building to work as a unit. That engineering makes a number of assumptions and contains supporting specifications. These specifications are used to select the various components that will be installed in the laboratory. The only way to know if these components are working together as specified is to perform the appropriate testing. This type of original testing is often referred to as commissioning.

There are few government regulations requiring fume hood testing. In biology and life science laboratories a device similar to a chemical fume hood known as a biological safety cabinet (BSC) is used to contain hazardous biological materials. The contaminant is biological rather than chemical. These BSCs are required to be inspected and tested annually, there is not a similar requirement for chemical fume hoods, even though a chemical fume hood is much more likely to have loss of containment than a BSC.

There are three different types of tests based on when and where the fume hood is tested. The first type of test is the AM or "as manufactured." This is generally a test that is done in a special testing laboratory. It is a test used to demonstrate that a fume hood could operate safely. The test laboratory appropriately creates a nearly perfect set of room conditions. This is because the AM inquiry is a test of the fume hood, not the room. To assess a fume hood's performance capabilities, you need to eliminate any impact caused by room conditions. Just because a fume hood can perform at a high level doesn't mean that it will. It is just one of many components in a very complex mechanical system.

The next type of test is the AI or "as installed." In this test, the fume hood's performance is observed once the hood is connected to the LVS. This is sometimes called commissioning and is the first indication that the fume hood is working as specified in the construction documents. This is not testing the fume hood independently, but rather testing how the hood performs when connected to the mechanical air system.

The final type of test is the AU or "as used." This test evaluates the performance level once the fume hood is loaded and a person is working at it. Again, you are testing the fume hood's performance along with the performance of the mechanical air system with the added factor of the fume hood being loaded as it would be in actual use with an operator in front of it.

These three types of tests can be performed in a number of ways to a number of different standards. Each type of test has a purpose and all three tests should be performed. Until the market provides new technology that gives the user a better indication of real-time performance, the only way to know if a fume hood is performing safely is to test it. Unless you test, you don't know if the hood is working or not.

When purchasing a fume hood, how can you tell if brand "X" is a better design than brand "Z"? It is very hard to know which product will perform best because the fume hood is not a standalone device. When a specialized test laboratory for fume hoods is built, the guidelines in the various standards are followed and the goal is to build the perfect room environment. This is done for two reasons. The first is to standardize test results. This standardization allows a purchaser to know each product was tested under the same test conditions so they can make equal comparisons between brands. The second reason for requiring the testing laboratory to be built to a consistent standard is to test only the fume hood. Testing all fume hoods in an equivalent environment makes the only variable affecting performance the fume hood itself.

Many fume hoods are designed without the benefit of having a fume hood test laboratory that meets these standards to test and optimize the design. If these fume hoods work well, its just because of luck, not science. If you want to design a world class fume hood, you need a world class test laboratory to test it in. That is why it is important for every company that is designing fume hoods to have a good testing laboratory. When purchasing a fume hood, you should ask about the supplier's test laboratory.

Why is designing a really good fume hood so difficult? The reason is that it is difficult to visualize and quantify what is happening within the hood. Years ago, there was a meeting of some of the world's best fume hood designers. The purpose of the meeting was to develop a new fume hood standard. As often is the case when you get a group of experts together, there is a lot of conversation. During this meeting, a very well-known expert said, "We don't design fume hoods for maximum performance, we design fume hoods that will test well." It is important to realize that each of these testing standards is evaluating something slightly different. Some fume hoods will test well regardless of how they are tested, but other hoods will test well on one standard and not so well on another. Why is this the case? Because the designers of those hoods were focused on optimizing the performance around a certain standard. Having tested many hoods using a number of different standards and tests, I would suggest that the more standards you test to (at the AM level) the better your chances of designing a highly effective fume hood. The more ways we challenge a fume hood in the test laboratory, the more robust it will be in the real world.

A well-designed fume hood that performs well in the test laboratory may not perform that well once installed in a real laboratory. The reason again is that the hood requires a mechanical air system to function. When AI testing begins, both the fume hood and the laboratory ventilation system together are being evaluated as a unit. A great hood will show a lesser performance when connected to a poorly designed LVS and a poorly designed fume hood might perform acceptably when connected to a great LVS. The ultimate performance will be a combination of the fume hood and LVS together. The logical first step is to acquire the best fume hood you can afford. The best fume hood is the one that has been tested to several of the standards and performs exceptionally well on them all. Starting with a fume hood that has demonstrated very robust AM performance provides the best chance of having an installed hood that performs well under a wide range of room conditions.

Once the LVS has been designed and the fume hood has been selected, AI testing will determine that the fume hood and the LVS perform together effectively and to the design specifications. Testing at this point holds the suppliers accountable for fume hood performance and failure to perform has to be corrected at the supplier's expense. Having a project commissioned is the only logical way to make sure the fume hood will perform as expected with a particular LVS. A great fume hood may not perform to specification because of problems with the LVS. Another reason to have an AM test for every hood model being installed, is that if the AI test results are poor, a comparison with the same AM test can reveal whether the problem lies with the fume hood or with the LVS. How can such a comparison reveal that? The reason is that the hood design didn't change, the poor performance is an indication that the LVS is the problem.

AI testing provides evidence that the entire mechanical system is functioning as designed and becomes a benchmark for future testing. AI testing is done when all mechanical systems are installed and the systems have been calibrated and balanced. This is done before the laboratory is occupied. Fume hoods are tested one at a time with no equipment in the hood. So, while this a very important test and should be done, AI testing is somewhat static and not reflective of the actual conditions in a working laboratory.

The next step in the life of a laboratory is actual operation. As the laboratory becomes operational, fume hoods are set up and used, people are moving around, doors open and close, sashes open and close. Building systems attempt to keep everything in balance. This is a much different environment from what was tested in the AI tests. If an AU test is performed and a fume hood tests poorly, unless AI testing was also performed, there is no baseline to determine the cause of the poor performance. Each test (AM, AI, and AU) builds on the previous test.

When a fume hood tests poorly in an AU test, you must ask, "Is the fume hood testing poorly because the LVS system has issues, or does it have more to do with the equipment setup or the work practices?" Once it is understood when and why to test, it's time to consider the various testing methods. There are many testing standards around the world. They range from very good indicators of fume hood performance to almost useless. Globally, two of the most commonly used standards are the ASHRAE 110 (current version at time of writing is 2016) and the EN 14175 (current version at time of writing was 2019). ASHRAE (American Society of Heating, Refrigerating and Air-Conditioning Engineers) has produced an American standard. ASHRAE is a worldwide professional association with members in 132 countries and it publishes many standards used worldwide. The EN (European Norm) is a European standard. It takes related standards from various European countries and normalizes them for use in the European Union. The EN standards draw from the old British Standard and the German Standard. For years there has been a debate about whether ASHRAE or EU standards are better. The real question isn't which is better, it is what is each of them testing and how are they testing? The test methods are very different. They use different methods to measure different functions. In the end they both measure some form of containment, but in very different ways. I suggest that fume hoods be tested by both standards because the challenges are different. Performing well by both standards is an indication that you have a robust product.

ASHRAE 110 is not a performance standard, but rather a testing protocol. You can't pass or fail ASHRAE 110 because it does not include performance criteria. ASHRAE 110 is generally used with the performance standards set in the ANSI/AIHA/ASSE Z9.5 Laboratory Ventilation Standard to determine acceptable performance criteria.

ASHRAE 110 focuses on the standard bench mounted fume hood. If a different type or style of fume hood is used, the testing methods must be modified. In the standard's appendix section, there is information that will help you decide how the test should be modified, but these appendices are not a required part of the standard and are provided for reference only. There is also a lot of talk about the use of the words "Shall" and "Should" in a standard, shall is a requirement while should is a good practice but not required. Today's best standards are a combination of education and requirements. While they describe the minimum performance requirements, they also discuss best practices.

One of the challenges is that if you look worldwide, you will find well over a dozen standards that apply to laboratory safety, laboratory ventilation, fire codes, and fume hoods. So even for someone in the business, it is quite a challenge to stay current with all these documents. In China there are several groups trying to write a comprehensive standard that consolidates all the issues for all the stakeholders into a single document. If successful, this could be the first attempt ever to create a comprehensive standard that addresses all the issues.

ASHRAE 110 and EN 14175 both test containment, but do so in very different ways. A very good fume hood will likely perform well when tested to both standards; however, there are cases where a specific hood will perform well on one test and not so well on the other. This most likely results from a fume hood being designed using just one of the test methods for guidance. When a manufacturer is designing fume hoods, the best practice is to use a number of the different tests to make the hood as robust as possible. Similarly, a fume hood designed and tested in a superior test laboratory is likely to perform better than one designed and tested in an average laboratory. The better the test laboratory, the better the resulting fume hood.

The reality is that neither test is perfect and neither test provides a definitive answer about how a fume hood might perform in a real-world situation, any testing is better than no testing. If everyone tests to the same standards under the same conditions, it does provide a valid, objective method of comparison.

What are some of the differences between the two tests? Currently, they both use sulfur hexafluoride (SF6) as a tracer gas and a leak detector to measure loss of containment.

The EN 14175, Part 3 test uses a 10% SF6 – nitrogen mixture with a total delivery rate of 2 liters per minute (lpm). there is an interior plane test which uses a 9-point diffuser sampling array. for the exterior plane and robustness test there is a release rate of 4.5 LPM. These challenge rates (CRS) are much smaller than the challenge rate of 4 LPM of 100% SF6 used in the ASHRAE 110 test.

With ASHRAE 110 not only do you have a more concentrated amount of tracer gas being released, but the instrument requirements are less and the data recording requirements are less. ASHRAE 110 is a series of highly localized challenges, whereas EN 14175 uses a grid system of averaged readings. With the EN testing, if there is leakage, it is difficult to troubleshoot. Some of the testing experts claim that the Containment Factor is not suitable as an objective indication of fume hood containment unless all models in question are tested under the same test conditions with the same volumes and the same leak detector equipment.

While the EN tests may show the ability of a fume hood to contain under certain conditions, ASHRAE 110 tests produce more finite readings that all fume hood designers can use to fine tune their designs, yielding more robust performance overall.

Both tests have unique strengths and weaknesses, because they are so different, it is challenging to make a determination about the fume hood's actual performance when comparing the results of the two tests. It would be hard to conclude that one test is better than the other because of the radically different approaches. The more types of testing conducted on a fume hood, the more clearly the hood's strengths and weaknesses can be understood. The best way to know how a fume hood will function is to install one in your laboratory with a typical setup and conduct AU testing.

There are several other world standards for fume hood testing. If you are in a part of the world that uses something other than ASHRAE 110 or EN 14175, use of those standards, in addition to these international standards, is encouraged. Every different testing method will yield different data and insights to a fume hood's potential performance.

There are three other tests which are modified versions of the ASHRAE 110 test. The first is Human As Mannequin (HAM) testing. In this test, a person straps on a leak detector, usually the Q200, puts the probe in their breathing zone. The person then performs a choreographed set of movements in and around the fume hood. This test is very dynamic and much closer to how a fume hood will actually be used.

The second modified ASHRAE test is the National Institute of Health (NIH) Fume Hood Protocol. This starts as an ASHRAE 110 test and is modified by adding a load (simulated equipment) to the fume hood. The SF6 flowrate is increased from 4LPM to 6LPM during a period of 5 minutes. At the conclusion of the gas test, a person rapidly walks by the fume hood three times. Acceptable test results will be a loss of containment of 0.05ppm (parts per million) or less.

The next area of focus in the NIH protocol is on face velocity. For this test, three ultra-sensitive omni-directional probes are required. In grid pattern, measurements are taken with the sash in the opened (18-inch) position and in the closed (6-inch) position. This process yields a measured face velocity. Then the steady state face velocity is measured. This will show how stable the face velocity is and the amount of fluctuation. Last, the turbulence intensity is measured. Collectively, these readings can reveal the stability of the face velocity which indicates the ability of the fume hood to contain hazards.

The NIH is a pass/fail test, and requiring containment of 0.05 ppm or less as an acceptable rate is a much more difficult challenge than the standard ASHRAE 110 tests.

The U.S. Environmental Protection Agency (EPA) produced a test called Performance Requirements for Laboratory Fume Hoods. This test focuses more on the low flow high-performance fume hoods. The EPA starts with ASHRAE 110, adds the NIH simulated load to the fume hood, and tests at various sash positions and face velocities. Again, this is a much more challenging test than the standard ASHRAE 110. The test is performed with the sash fully open, then 80% open, 40% open, and finally closed (6 inches open). Each position is tested at a face velocity of 100 feet per minute (fmp), 80 fpm and 60 fpm. At the 6-inch opening, the face velocity cannot exceed 300 fpm. The hood static pressure must be below 0.25 wg. (inches of water on a water gauge) The tracer gas must have a 5-minute average at or below 0.05 ppm and a maximum 30 second rolling average at or below 0.10 ppm.

There are many ways to modify the ASHRAE 110 to be more realistic for your situation. Unfortunately, this type of testing is somewhat expensive and is only a snapshot in time. While AM, AI, and AU should be done on every hood initially, what about follow-up testing? These tests are qualitative tests that are standardized with numerical data that is repeatable, but there is also qualitative testing which is more observational. This type of testing allows you to visualize what is happening and many of these tests you can do yourself. One of the most basic observational tests involves pouring warm water onto dry ice in a dish in the fume hood. The resulting fog will provide some idea of how the air is moving inside the hood. Theatrical smoke generated by a fog or smoke machine is another good way to visualize air patterns and movement. To further enhance this, pointing a handheld laser modified to have a flat beam into the smoke amplifies the visual observation of the turbulence.

There is also a new technology soon to be on the market called Predictive Containment which uses sensors and artificial intelligence (machine learning) to monitor a fume hood's containment in real time. These "smart" hoods will be able to monitor the hood's containment continuously.

Another real-time way to test containment is to monitor the air quality just outside the fume hood. This concept acknowledges that the fume hood will likely have some loss of containment. When loss of containment occurs, this air monitoring system will detect fugitive chemicals and increase the number of ACH to dilute these contaminants to a safe level.

Lastly, it should be noted, that because SF6 has been identified as harmful to the atmosphere (greenhouse gas), there is a movement underway to change the tests that use SF6. The most talked about option is to change from SF6 gas to isopropyl alcohol (IPA).

The next few years will likely bring much change to the world of fume hood testing, including a more global requirement to perform regular testing, for now, any testing is better than no testing.

Chapter 12

Hood Design – What Makes A Great Fume Hood?

A fume hood's design makes a huge difference in performance. The more robust the design, the better the fume hood will perform, even under adverse conditions, but given the fact that the fume hood is dependent on the laboratory ventilation system to perform properly, why is design important? Laboratories are very complex structures making it difficult to manage airflow and balance. Laboratory room conditions create challenges for all hoods. The better the fume hood, the better it will perform under all these ever-changing conditions. I have tested thousands of fume hoods and observed even more. Some hoods are so poorly designed that you can tell, just by looking at the fume hood, that it will not work properly. Some designers simply copy features they have observed on other fume hoods without understanding the significance of the details of the features. These copies often don't perform as intended. Many fume hoods are designed by furniture companies that have designed the fume hood as if it were a piece of furniture, when it is actually a very complex mechanical device. Because air and water follow the same physical laws (fluid dynamics), a fume hood is more like a pump than a piece of furniture. If you designed a pump to move a liter of liquid per minute, to increase the volume to 10 liters per minute, you don't simply make the pump bigger. Yet with fume hoods that is often the approach. Fume hoods have a three-dimensional airflow and the airflow isn't linear; as the dimensions change so do the airflow patterns. Even the same model of fume hood in the different sizes will perform differently.

Why is it that fume hoods are most often designed by furniture companies when fume hoods are a significant part of the mechanical laboratory system? Wouldn't it make more sense if fume hoods were designed by the people designing the other mechanical components? The answer is deeply rooted in history. Most people trace the first modern fume hood back to the University of Leeds (UK) in 1923. It can be described as taking a wall cupboard and mounting it over a window. That is why in Europe the fume hood is called a fume cupboard. Much of the early fume hood development was done prior to 1950 by furniture companies that included these devices as part of their furniture package. Because the fume hood is the interface between the user and the mechanical air system and is located in the laboratory space, architects and laboratory planners wanted fume hoods to match the furniture. The tradition of fume hood design by furniture makers has continued without challenge.

In order to design a great fume hood, it is necessary to understand all parts of the fume hood, the LVS and how all the various parts work together. The significance of airflow in achieving containment requires a designer to understand that very small details can have a big impact on performance. Recently, I visited a university laboratory where they were able to demonstrate that a 50nm (nanometer) strand stretched across an airflow causes turbulence. Something as simple as a screwhead in a flow path causes air turbulence and eddies. The perfect design would create laminar flow with minimal turbulence. A fume hood is not a static device, the sash opens and closes and the load inside the fume chamber changes. People interface with the hood putting things in and taking things out. All this dynamic change creates unlimited operational variables. Now when it is put it in a laboratory with other fume hoods, more people and varying room conditions, you have infinite challenges to overcome. As the situation becomes more complex, it is difficult to maintain safe operation.

Fluid flow is often chaotic. A measurement known as the Reynolds number (Re) is used to help predict flow patterns in different fluid flow situations. The Re is calculated by determining the ratio of inertial forces to viscous forces in a liquid and is used to predict if a flow condition will be laminar or turbulent. Laminar flow (or streamline flow) occurs only at low Re numbers below 2000 when a fluid flows in parallel layers, with no disruption between the layers. At low velocities the fluid flows without mixing and adjacent layers slide past one another like playing cards. In laminar flow the fluid motion is very orderly with all particles moving in straight lines parallel to the vessel walls.

Turbulent airflow occurs at higher Re numbers above 4000 and is dominated by inertial forces, which produce random eddies, vortices and other flow fluctuations. Unless a vessel, such as the fume hood, is designed to control the pattern of flow, turbulence occurs. The zone between laminar 2000 Re and turbulent 4000 Re is a transitional zone that could be either laminar or turbulent depending on conditions within the vessel. Re numbers predict conditions where airflow is in constant motion to a fume hood's internal surfaces. For airflow moving in a rectangular shape such as a fume hood (sash face opening or upper chamber), an equivalent diameter is calculated.

Calculating the fume hood's Re number will either validate or debunk every "rule of thumb" myth surrounding fume hoods. When a fume hood's liner becomes dirty the fume hood's performance will deteriorate. A dirty hood will have an increased Re number and increased turbulence. A longer length fume hood will produce worse performance than a shorter length fume hood. Increasing face velocity beyond 120 FMP degrades performance. The following graph illustrates different operating Re numbers between a four-foot and eight-foot standard bench hood at either 50 fpm or 100 fpm face velocity. You can see the dramatic change in Re numbers as face velocity and hood length change.

Re Number		
38,000	8 foot hood @100 fpm	
36,000		
34,000		
32,000		
30,000	4 foot hood @ 100 fpm	
28,000		
26,000	8 foot @ 100 fpm	
24,000		
22,000	4 foot hood @ 100 fpm	
20,000	8 foot hood @ 50 fmp	
18,000		
14,000	4 foot hood @ 50 fmp	
12,000	8 foot hood @ 50 fmp	
10,000	4 foot hood @ 50 fmp	
8,000		
6,000		
4,000	Above 4000 is Turbulent	
	Transitional Region	
2,000	Below 2000 is Laminar	
	18" Sash Opening	28" Sash Opening

Fume hoods are very

turbulent, even at reduced face velocities. The reason some fume hoods perform better than others is solely dependent on their internal design. Conventional narrow depth "Walk-in" floor mounted fume hoods perform poorly because their Re numbers are double that of bench hoods, indicating twice the turbulence.

An error made by fume hood designers and industrial hygienist was to assume that because the face velocity was fairly uniform across the sash chamber opening, then so too were the airflow patterns inside the fume hood.

The airflow entering an enclosure will have a laminar profile but quickly become turbulent. This is exactly how a fume hood operates.

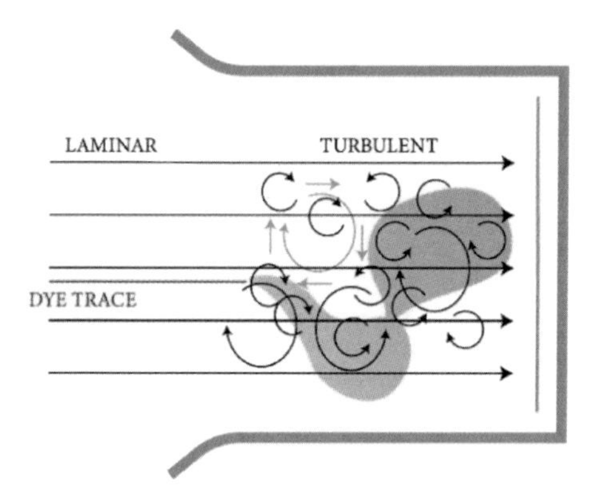

Users, as a critical part of the system, should have enough knowledge about the fume hood they are using to make informed decisions about fume hood safety. Their actions will impact the fume hood's ability to perform properly.

Size matters. The airflow in a fume hood is three dimensional. Because the Re number is affected by dimension, a 48" hood, a 60" hood, a 72" hood, and a 96" hood of the same model, from the same manufacturer, will have different performance results. This is because the airflow patterns are not the same. Every change to the hood has an impact on performance. Some changes make the fume hood better while others reduce performance. It is the minor details that separate a good fume hood from a great fume hood.

Where do the typical dimensions of a fume hood come from? Because fume hoods are most often designed and built by furniture manufacturers, the dimensions of fume hoods have been unduly influenced by the size of doors and ceiling height and the design of the laboratory spaces. The fact that a fume hood is aesthetically pleasing and fits through the door and onto a laboratory counter doesn't mean it will perform well. The dimensions of the standard hood don't necessarily optimize its performance.

Air is lazy, it will follow the path of least resistance. The good news is that airflow is manageable. The airflow inside the hood is three dimensional with a horizontal and a vertical vector component, and as these components collide, it produces air turbulence. Much like water, as the air flows across various physical features, the air is curved and bent and the speed and direction change. What might have started out as a smooth even flow can quickly become very turbulent.

If the fume hood stayed in the same state of operation and the setup never changed, we could easily optimize the airflow for great performance. Somewhat like automobiles or even bicycles, if they were only driven in the city at slow speeds then they could be optimized for that. As soon as the driver takes them off-road, everything changes. In the case of a fume hood, we have a sash that is always being opened and closed, the setup is changing and more importantly the room conditions are changing. Given the dynamic nature of fume hood operation, the product needs to be as robust as possible to provide the best chance for proper performance over this wide range of conditions.

After the size of the fume hood has been determined, what should a designer consider? First, how the air enters the hood. There are three surfaces to consider. The single most important surface is the lower airfoil. Next is the sash handle and then the vertical components which frame the sash opening. These are the left and right front posts, known by many names including columns, posts and facia. Each of these surfaces have an impact on how the air enters the fume hood.

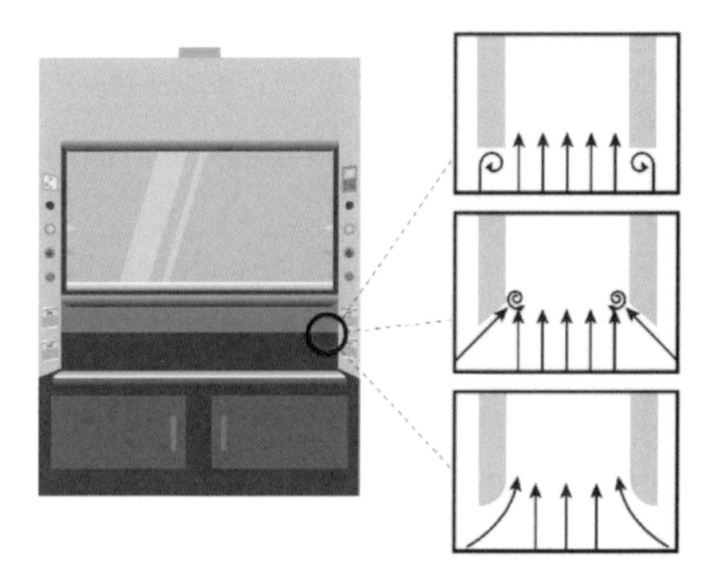

Next, what happens inside the fume hood? The air is pulled towards the back baffles because of the lower differential pressure, but this movement is not uniform. If you take a velocity probe and read the slot velocity across the back at the bottom of the hood, you will find it varies from left to right. You can do that at any other slot or horizontal opening and see many different readings. This uneven suction across the baffles causes the air to follow the easiest path. The exhaust duct is usually in the center of the top and the suction is stronger in the center. So more of the air entering the hood is pulled towards the center. Openings, or slots, near the top of the hood will have stronger suction than those near the bottom. Even if the airflow entering the hood was uniform, the further into the hood it goes, the more turbulent it becomes. The art of designing an efficient fume hood must take into consideration the placement and design of the baffles and their openings, and the pressurization of the plenum created by the baffle.

There are fume hoods on the market and hoods in development that have made improvements in balancing the flow going into the baffles. Some fume hoods actually have created a laminar flow where the air enters the sash opening and flows directly to the back of the hood remaining very laminar. Other designs have achieved a laminar flow at the top of the hood above the sash opening eliminating or greatly minimizing the traditional upper vortex.

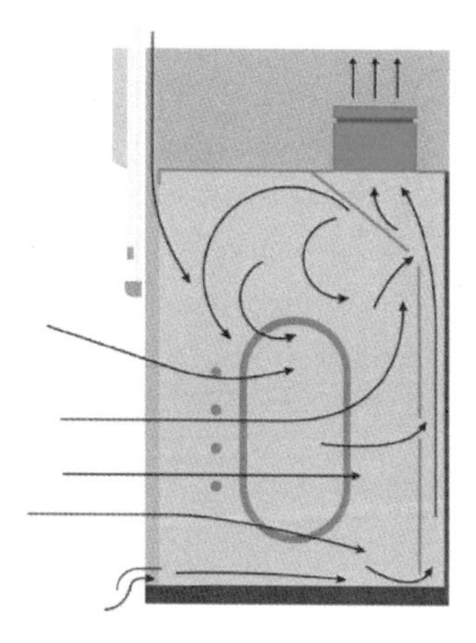

The more organized the airflow, the better the hood will contain contaminants, regardless of face velocity. The less turbulence in the fume hood, the better the containment. The best hoods are designed to be very aerodynamic with reduced turbulence and organized flow, but the only way to really know how well a hood performs, is to test it. As discussed in Chapter 11 on testing, some hoods test well using one testing standard, but not so well using another. The more standards or real-life situations you test against, the better you will understand the performance of your hood. Any visualization aids you can use will help you better understand what that invisible air is doing.

The majority of fume hood designs have a vortex at the top of the hood. This vortex is caused by air coming in the open sash and going to the upper back of the hood. This allows the air in the front top area of the fume chamber to just spin creating a vortex. As the sash opens and closes the vortex usually moves up and down with the sash. If the sash has been closed, the vortex has likely dropped nearly to the worktop. If the sash is opened quickly, it is possible for that buildup of contaminants to spill out before the hood can stabilize potentially exposing the user. Designs that either eliminate the vortex or prevent it from dropping down when the sash is closed will perform better and have a higher containment ability.

One of the last points worthy of discussion regarding fume design is the sash style. A single vertical rising sash will always perform better, but, horizontal sashes and combinations of vertical and horizontal sashes are also being used. Why do vertical sashes outperform the other styles? When you open only one side of the sash, which is common with horizontal panels, you create a second vortex. This vortex is vertical. This can create a rotation in the flow between the back of the hood and the sash opening. Contaminants are collected deep inside the hood and are brought up to the face of the hood and into the user's breathing zone. There are a lot of misconceptions and misinformation touting that horizontal sashes and combination sashes are safer, but the evidence shows otherwise.

In the end, a great fume hood is one that provides safety under a wide range of operating situations. It is robust and able to compensate for less than ideal room conditions. Remember the size of a fume hood has a big impact on performance and that minor design details can be the difference between a good fume hood and a great fume hood.

Chapter 13

Classes of Hoods and Risk Management

It is not always necessary to use a high-performance fume hood. Taking a risk-based assessment approach to determining the needs of each particular laboratory will produce a much more efficient and cost-effective safety plan. The one-size-fits-all approach we have used for years is not the optimal solution. Whereas one laboratory may require a top of the line, high-performance fume hood and a complex LVS to maintain a safe environment, the next laboratory may need a far less sophisticated system to achieve the same level of safety. Even within the same building, risks can be very different from one area to another.

The American Society of Heating, Refrigerating and Air-Conditioning Engineers (ASHRAE) and the Scientific Equipment and Furniture Association (SEFA) have both embraced the risk-based approach to laboratory safety, but in different ways.

SEFA's Approach

SEFA released a document titled, *Guide to Selection and Management of Exposure Control Devices in Laboratories.*

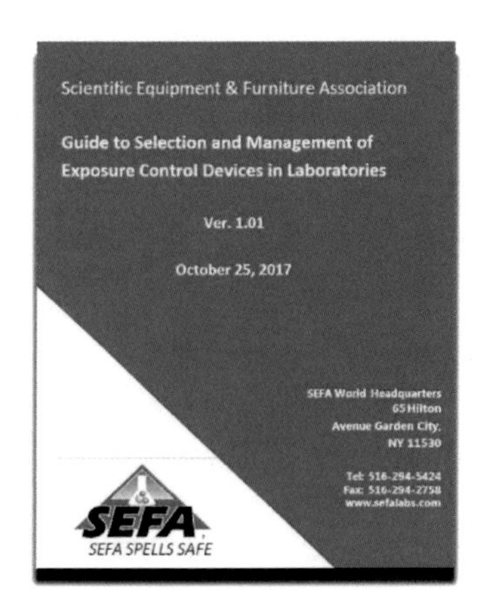

Scientific Equipment & Furniture Association

Guide to Selection and Management of Exposure Control Devices in Laboratories

Ver. 1.01

October 25, 2017

SEFA World Headquarters
65 Hilton
Avenue Garden City,
NY 11530

Tel: 516-294-5424
Fax: 516-294-2758
www.sefalabs.com

SEFA
SEFA SPELLS SAFE

This document is a reference manual that discusses the risk of user exposure in the laboratory. Based on the risk in a particular laboratory, different types of Exposure Control Devices (ECD) are recommended. A traditional fume hood is a class of ECD. This Guide seeks to strike a balance between safety and energy usage based on risk to the user. The underlying message is that all risks are not the same and therefore the optimal solution for different risk levels varies.

Fume Hoods are only one group within the exposure control device family. Remember that not everything that looks like a fume hood is a fume hood and not all fume hoods perform the same.

At the lowest level of the ECD matrix, is localized exhaust. This can be snorkels, or spot extractors, or canopy hoods. Often these devices are used to remove heat as much as chemical fumes and are designed to be close to the source. In the next level of the ECD matrix, is a class of products often referred to as a ventilated enclosure. This class covers a wide range of products. These products include vented chemical storage cabinets, machine enclosures, balance enclosures, small ventilated rooms, and cylinder cabinets. Ventilated enclosures also include devices such as downdraft tables, slot hoods, and double-sided demonstration hoods. Virtually any device that is vented but that does not meet the criteria and performance to be a chemical fume hood is considered a ventilated enclosure.

Not all exposure control devices are vented, and not every ECD can be effectively used with certain materials. Close attention must be paid to the application and the intended use of the ECD.

Biological safety and chemical safety have a similar objective and that is to protect the users from exposure. Biological safety is more defined and regulated than chemical safety. The primary reason is that the effects of biological exposure are more immediate. Sickness or symptoms of biological exposure may appear much sooner than the effects of chemical exposure. The effects of chemical exposure are often not apparent for years.

In the biological world, safety levels have been defined based on risk for years. Biological Safety Levels are defined in the United States as BSL1, BLS2, BLS3, BLS4. The BSL1 calls for simple protection since the risk of exposure to a harmful agent is low, but at BSL4, users must employ sophisticated protection methods to handle the deadliest agents. In a life science laboratory, the equivalent of a fume hood is a Biological Safety Cabinet (BSC).

These devices are used to protect users from biological hazards. They look somewhat like a fume hood, but operate in an entirely different way. Their performance is rated by class. But the most important thing to understand here is that these are not chemical fume hoods and should not be used as if they were.

BSCs are divided into classes that offer different levels of protection based on the type of work being done and the risk of exposure. BSCs have their own definitions and set of requirements for use. In general, these devices are standalone and are not connected to the laboratory ventilation system. Their air is generally filtered and recirculated. BSCs are designed and intended to be used with biological materials only -- not chemicals. The general rule is, do not use a chemical fume hood for biological work and don't use a BSC for chemical work. While they might look similar, they have entirely different functions and performance.

In the chemical world there are also a number of types and classes of fume hoods. But unlike BSCs, they are not so well defined or categorized into groups or classes. Here are a few of the hood types used in chemical laboratories:

Ductless Hoods

Ductless hoods are more similar to a BSC than a chemical fume hood but are intended for limited chemical use. Ductless hoods are filtered and they also recirculate the air. Depending on what chemicals are being used, these ductless hoods are outfitted with different types of filters.

Generally, these devices are intended for specific applications and not general use. If you are using a ductless hood, check with the manufacturer to verify that your application is appropriate for the device. Most manufacturers of these devices do an audit of the work planned for the laboratory before recommending use of a ductless hood.

Laminar Flow/Clean Bench (Horizontal Laminar Flow and Vertical Laminar Flow)

Laminar Flow/Clean Benches are generally used to protect the product inside the hood and not the operator. They are much more common in a cleanroom than in a chemical laboratory. These devices support very specific types of work. Often it is hard to tell exactly what these units are. They may look like a chemical fume hood, they may look like a BSC, or they may look like both. It is very important to know what type of device you are working with, what types of work the device was designed to perform and whether it is working properly.

Clean Air Hoods

Clean air hoods are strictly for product protection. They are not intended to protect the user in any way. Generally, they take room air, bring it through a HEPA filter, and blow it on the product. This creates positive pressure inside the hood rather than the normal negative pressure used for operator protection.

Chemical Fume Hoods

This is the class of products that this book focuses on. These devices are generally found in chemical laboratories or in places where chemicals are being handled and where the user could be exposed. A survey of the world market will reveal many brands of chemical fume hoods. In fact, it is not uncommon for a single manufacturer to have multiple brands or families of products. While the intent of all chemical fume hoods is the same, to protect the user from exposure, the approach used to accomplish this varies considerably. Also, the designs can be very different. Again, unlike the BSC and the Ductless Fume Hood, the Chemical Fume Hood requires connection to a laboratory ventilation system to work. The fume hood is merely the user's interface to the mechanical air system.

Traditional Bypass

The traditional bypass is one of the oldest types of fume hoods. It is a constant air volume hood. As with any hood, if the air volume being exhausted is constant, and the sash opening is reduced, the face velocity will increase. Without another way for air to enter the hood, the face velocity can increase to unsafe levels. At the top of a bypass hood there is another opening called a bypass. This bypass opens as the sash is closed and closes as the sash is opened. The total air coming into the hood between the sash opening and the bypass helps keep the face velocity at a safe level.

High Performance

The label "high performance" can be applied to a number of hood types. It is not really a class of its own. It generally means that the hood can perform safely at lower face velocities. Thus, achieving the same performance with less air. Hoods that carry this label are generally more robust, meaning that they have a better chance of performing well, even when the room conditions are not ideal. For example, if a high-performance hood tests well at 40 fpm under ideal conditions, it might do better than a standard hood at 80 fpm or 100 fpm when the overall laboratory ventilation system is weak.

VAV Restricted Bypass

This is a form of bypass hood intended for use with a VAV (Variable Air Volume) exhaust system. A VAV fume hood adjusts the volume of air automatically as the sash opens and closes. Because there is less air as the sash closes, the bypass is not needed for face velocity control, but because chemical build-up can occur within the hood when there is no air coming into the hood, the restricted bypass allows a small amount of air to enter the hood when the sash is fully closed.

Auxiliary Air

This is an older type of fume hood. It was one of the first approaches used to reduce the amount of room air used. An auxiliary air fume hood has a plenum at the top front of the hood. It discharges outside air in front of the sash so the hood will use that air instead of exclusively drawing air from the room. The main problem with this concept is that the outside air is not conditioned, so it can be hot or cold, dry or humid. Because this air is directed down on the user, it can be uncomfortable to work at this type of hood. Auxiliary air hoods are still being specified and used, but they are only suited to certain situations. Many will argue that Auxiliary Air hoods tend not to be safe because of the lack of control over the outside air being released in front of the hood.

Synchronized Supply

This is a one of the newest classes of fume hoods. There are a number of synchronized supply designs in the market and each design functions a bit differently. As discussed earlier, loss of containment can occur while a fume hood and the room are stabilizing from a shift in differential pressure. The primary purpose of this fume hood design is to control both exhaust and supply air within the fume chamber. By doing so, there is almost no lag time in air pressure balancing. The exhaust and the supply stay synchronized. With this concept, the hood is practically a standalone device. There is also a second advantage to this design. The source of the supply air can be outside air, unconditioned or partially conditioned or fully conditioned. There is a potential to save energy over using room air exclusively. Unlike an auxiliary air fume hood, the supply is discharged directly into the fume chamber in a non-turbulent stream away from the operator.

Distillation

A distillation fume hood is generally a traditional style hood except it is much taller. The reason for the additional height is to make it easier to work with the tall distillation columns. Sitting on a tradition laboratory bench, this hood would be too tall and would hit the ceiling. These hoods are usually mounted on a bench or stand that is only 12 to 18 inches off the floor. Where practical, the top of a bench hood, a floor mounted hood and a distillation hood are at the same level, often 84", for aesthetic reasons.

Floor Mounted

Floor mounted hoods are also known as a walk-in fume hoods. As chemical development moves from the bench size quantities to production quantities, there is a process called scale-up. The process often goes from the bench to a micro-factory that is housed in a floor mounted hood to a pilot plant and then to full production. The floor mounted fume hood is designed to hold larger equipment. These hoods range from small and simple to large and complex. As these floor mounted designs become larger it becomes more difficult for them to perform like a fume hood and they often become vented enclosures. While they look the same to the user, the performance can be very different. Floor mounted devices require a better understanding of the risks of the processes happening within them and require more testing to verify safe operation. The devices can become small rooms. Don't assume that all floor mounted hoods will function the same as a bench hood, because they will not.

Perchloric

This is a fume hood designed especially for use with perchloric acid and should not be used for anything else. Likewise, perchloric acid should not be used in any other type of fume hood. What makes this hood unique is the washdown system. As perchloric acid fumes dry into solids these crystals can become explosive. The water washdown system minimizes the buildup. Most perchloric hoods are made of stainless steel and have rounded corners for easy cleaning. They have a large trough in the rear to catch and drain the washdown water.

Radiation Hoods

Also known as a Radioisotope Hood, radiation fume hoods are usually stainless steel and often similar to a perchloric hood. The main difference is that there is no washdown system and the countertop is reinforced to support the large heavy lead bricks used in the fume hood. It is also very common for this style of hood to have a HEPA filter in the exhaust stream. Again, these are special fume hoods and should only be used with the intended materials.

High Heat Hoods

A high thermal load in the fume hood dramatically changes the airflow patterns. There are heat loads that range from small hot plates and open flames to steam baths and small furnaces. As the heat load increases, the normal performance of the fume hood decreases. The airflow design must be optimized based on the amount and location of the heat source. In addition to the changes in airflow, there may need to be special considerations related to the chemicals being used. For example, if you are performing acid digestion with hot hydrofluoric acid it will attack many materials, including glass. This is another case where it is essential that the specific application be considered to choose the appropriate fume hood.

Ultra-Low Flow Hoods

This class of product is designed to reduce the face velocity as much as possible. Often the design includes forced air being introduced into the hood chamber. Generally, these fume hoods are very aerodynamic in an attempt to maintain containment while significantly reducing the amount of air they consume. While many of these products are well designed, because of the low air velocities used, they tend not to be as robust as some of the other designs and generally work best with very well-designed and maintained laboratory ventilation systems.

Reduced Air Volume Hoods

This style fume hood usually has restricted sash movement. By restricting the sash opening, the hood uses a reduced amount of air. While this style product does exist, it is more of a marketing concept as opposed to a technical one. The limited sash opening may make the use of this product limited, and overriding the sash restrictor may make the hood unsafe.

The following designs achieve the next level of protection and are more contained.

Glove Boxes

These devices are generally totally enclosed and accessed through glove ports. They often have airlocks for moving samples in and out and use inert atmospheres inside such as nitrogen. It is important to remember that just like fume hoods, glove boxes are a whole class of products that have many different performance ranges for specific applications. Care should be taken to understand the capabilities of each type of glovebox when selecting one for your application.

Isolators

These devices often look like a glovebox. In fact, they often have many glovebox features but are used mostly in the pharmaceutical industry to meet aseptic and containment requirements. Typically, isolators are custom designed to fit the application.

As you can see from this list of ventilation and containment devices, an entire book could be written on the different types of devices and their intended uses. Even if such devices are never needed in your work, it is helpful to be aware of the different types of devices and how they work. The more you understand, the safer you will be working around these devices.

Not all exposure is the same. The more toxic the chemical, the greater the risk. A user that only uses a fume hood a few minutes a day has a different risk level than someone who works at a fume hood most of the day.

The established levels of biosafety recognize that there are different risk levels. Preventative actions taken for low risks (BSL-1) are much different than those taken for high risks (BSL-4). This is a very logical approach to risk management. On the chemical side, since 1991 OSHA has required each facility to have a CHP (Chemical Hygiene Plan), but these plans really don't do much to address the mechanical side of risk management. These (CHP) plans are more about process and procedure. In the late 1990s, when SEFA started rewriting SEFA 1 – Fume Hoods, I suggested that chemical laboratories should be assigned categories similar to BSL classifications based on risk. While there was little support for the concept at the time, finally 20 years later, ASHRAE has made risk-based classifications a new standard.

An old joke from Engineering school is that if you put your head in the oven and your feet in a refrigerator, your average temperature would be just right. This illustrates the problem with using averages in some circumstances. One size fits all is not an applicable concept where fume hoods are concerned.

ASHRAE's Approach

To address the issue of risk management as it relates to chemical fume hoods, ASHRAE has created a new standard that is risk based. They looked at the levels of exposure and defined a level of protection. This coupled with the SEFA ECD guidelines has finally provided a workable risk-based concept.

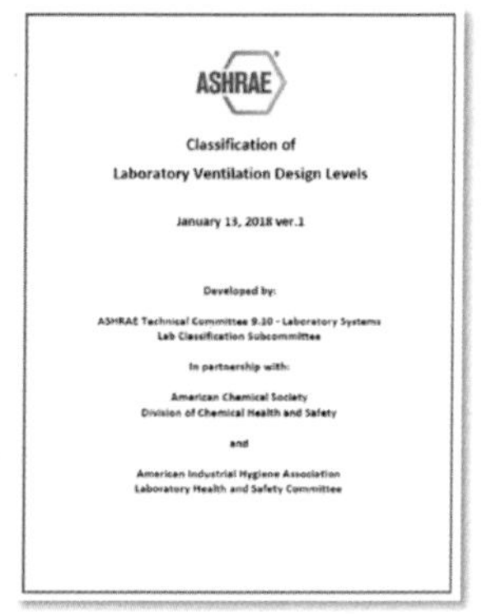
ASHRAE

Classification of
Laboratory Ventilation Design Levels

January 13, 2018 ver.1

Developed by:

ASHRAE Technical Committee 9.10 - Laboratory Systems
Lab Classification Subcommittee

In partnership with:

American Chemical Society
Division of Chemical Health and Safety

and

American Industrial Hygiene Association
Laboratory Health and Safety Committee

The use of risk-based design is long overdue and is a more logical way to balance safety and energy usage.

The ASHRAE Standard is called Classification of Laboratory Ventilation Design Levels. Much like the Biosafety levels, ASHRAE has established 5 ventilation system design levels. The levels are designated VSDL 0 through VSDL 4. The 5 levels correspond to risk levels from Negligible (0) to Extreme (4).

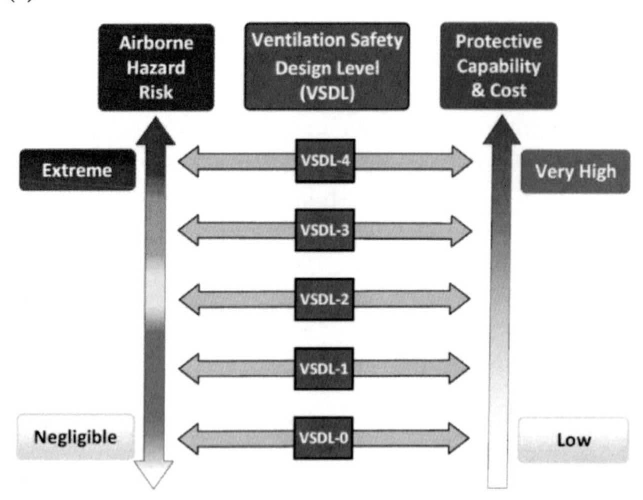

ASHRAE explains what types of chemicals and work are involved in each design level and gives performance criteria for each level.

While these classification standards can be implemented with fume hoods using today's technology, it offers some interesting future possibilities. As fume hood technology advances and fume hoods become more advanced, they could have the ability to adjust to different design levels. A fume hood could operate as a VSDL-2 this morning and switch to a VSDL-4 this afternoon.

This design level approach to fume hood performance seems to be the most adaptable to the current state of the industry and the technology currently in use. It also provides a platform for future innovation and the integration of "smart" ventilation systems.

For example, if the work I do in a certain laboratory is classified as VSDL-0, I really don't even need ventilation devices. The work can be done safely on the open countertop. However, if my work falls under the VSDL-4 classification, I must be much more concerned about safety and take many more precautions.

This ASHRAE safety level strategy gives us the opportunity to use multiple levels of safety precautions within the same building -- a zone concept. Instead of having too little protection in one area or unnecessary protection requirements in another, we can better match the risk of exposure to the level of protection we are providing.

In a VSDL-3 situation for example, the conditions are defined as having large quantities of hazardous materials and a high potential for airborne generation. The ASHRAE standard lists the following chemical GHS (Global Harmonized System) Codes as the types of chemicals that fall under VSDL-3 being chemical like H331, H332, H333, H334 & H335, H336 and H226, H227, 229, H242, H261.

The ASHRAE standard for VSDL-3 recommends 6 to 8 ACH (air changes per hour) while the laboratory is occupied and a reduction to 4 ACH when the laboratory is unoccupied. The standards additionally recommend dispersion modeling to assess building re-entrainment. Use of fume hoods that maintain constant air volume (CAV) should be considered when designing equipment for level 3 risks. In addition, differential pressure should be controlled and monitored during normal operation and continuous monitoring of laboratory exhaust airflow should occur.

In all, ASHRAE has identified 32 Laboratory Ventilation Design Criterion and established requirements for each of the 5 risk levels. When used in conjunction with the Laboratory Ventilation Standard, ANSI/AIHA Z9.5, the ASHRAE laboratory ventilation design classification system provides the most comprehensive requirements available.

As we transition from the one size fits all concept of exposure control to a risk-based concept for exposure control, there is a huge opportunity to not only better protect users, but to do so with less energy usage. As we approach one hundred years of modern fume hood use, we should pause and rethink what we are doing. Legacy ideas are hard to change, but today we have newer technology and a much better understanding of the risks involved. Now is the perfect time to evaluate our position on these devices and move forward in a way that is safe and sustainable.

Chapter 14

The Future of Fume Hoods

Historically, we have seen heating, ventilation and air conditioning (HVAC) technology change, sensors have improved, incremental improvements have been made in fume hoods. The fume hood industry has gone through periods of change and improved performance; however, the fundamentals of fume hoods have changed very little. The current fume hood design is about 100 years old.

The problems associated with fume hood performance have been long known, but the solutions have proven to be elusive. Fume hood designers, in the current environment must accept certain realities. First and foremost is that the fume hood is simply a component in an overall mechanical system. As much as we would like fume hoods to be standalone products, they simply are not. The best fume hood in the world does nothing until it is connected to a ventilation system. While a more aerodynamic fume hood can be designed, these improvements alone cannot fix the problems with fume hoods.

What are the problems with fume hood performance? The short answer is that fume hood performance is directly related to how effectively air volume/air pressure is managed. Airflow in buildings is very complex and managing that airflow is equally as complex. A superior laboratory ventilation system requires that the building's HVAC system be equally superior.

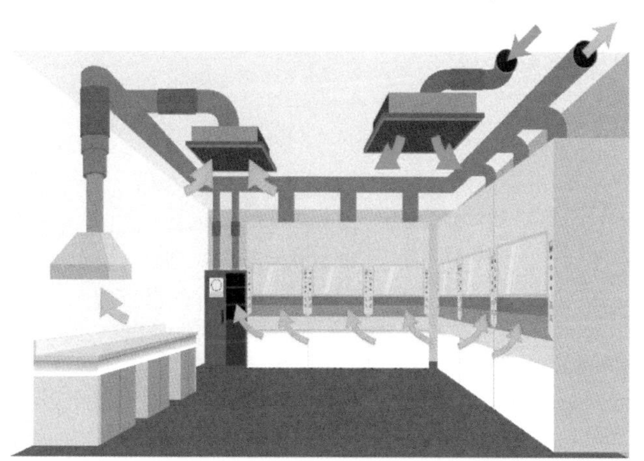

Let's examine a simple laboratory with six fume hoods on a variable air volume (VAV) system.

Assume that each fume hood uses between 150 CFM (closed) and 600 CFM (open)of airflow depending on sash position. When all fume hood sashes are closed, there would only be 900 CFM exhausted, but if all six hoods are being used with sashes in the operating position, they would be exhausting 3600 CFM. To keep the room properly balanced, when we go from all closed to all open, we need an extra 2500 CFM of supply air.

The room air supply must adjust to provide that extra 2500 CFM of airflow. How does the laboratory ventilation system know that more air is needed and how much? It depends on how the laboratory ventilation system is designed. Depending on the system, this rebalancing of airflow could take from a few seconds to a few minutes.

What happens in the laboratory while the system readjusts the air supply and the pressure becomes stable again affects a fume hood's performance. For instance, if one operator completes a task, the operator closes the sash on the fume hood being used. Immediately a totally different amount of airflow is required to maintain containment in the fume hood. If the laboratory door is opened, because the laboratory is under slight negative pressure, the air from outside rushes into the laboratory seeking to establish equilibrium. Again, the airflow requirement in the laboratory changes. How does this affect the amount of air needed and how long does it take for the laboratory to regain balance and stability? What is happening to the fume hood performance during this time? The fume hood can only contain contaminants when it is under negative pressure compared to the room. There is a high probability that during these periods of airflow adjustment, the hoods(s) have had a loss of containment.

During the course of a day, the airflow requirements constantly change based on the number of ventilation devices in use, how much airflow each one requires, as well as the number of times people go in and out of the laboratory. The laboratory ventilation system is always seeking air pressure balance. The complexity of the situation is obvious. Like that electrical outlet discussed earlier, fume hoods are the user interface to a much more complex system. The function of both fume hoods and electrical outlets depends on the supporting system and how the user interacts with the fume hoods.

The future is about making the fume hood smarter and a more active participant in the operation of the laboratory ventilation system. Fume hoods are currently a slave to the laboratory ventilation system. In the future it is desirable to establish a more peer to peer relationship between the fume hood and the ventilation system or making the fume hood an independent standalone device. There are many approaches being considered and many groups investigating new solutions to some of the age-old problems.

This is a good time to say that we have challenges to innovation. Legacy has been a strong influence over our current situation. Legacy is well rooted in government regulations, building codes and product standards. In the western world, the entire laboratory ventilation system is so regulated that it is unlikely that much true innovation will happen. However, as we go to Asia there is substantially less regulation and legacy. Although much of what has historically happened in this part of the world is to simply replicate what is being done in the west, because the roots of legacy are not so deep and there are fewer regulations and standards, more innovation is expected from Asia. They are already trying new concepts and proving these concepts work. As these technologies mature, Asian designs will be used by multi-national companies. Eventually, these concepts will make their way west. Look for the next major advances to come from the east not the west.

Now back to smart devices. In Chapter 3 there is a discussion about the electrical outlet and how it is the interface between the electrical system and the user. The standard outlet does very little outside the designed function of accepting electrical devices to be powered. But what happens when you make that outlet "smart" by adding additional functions to it?

While a smart plug is still the interface to a complex system, when coupled with a brain or control module, it has much more function. It can report its status; it can be controlled remotely. It can be triggered by other devices or an automation schedule. It can even be smarter and monitor devices plugged into it. It can collect data on usage and report it to a control system.

While the first smart fume hoods will most likely not be totally independent devices, they will have more control over their performance. Additionally, they will have more features that not only make them safer, but will also make them more convenient to use.

During the past 5 years, I have worked with a number of different organizations on ways to better integrate the fume hood and the laboratory ventilation system. There are many ways to approach making the fume hood smarter, but for the near future, the fume hood will always be part of the mechanical air supply system rather than a standalone device.

Yes, there are ductless fume hoods that filter the exhaust air and return it to the room, but these are most suited for limited applications and not general use. There is a place for this class of product, but it requires knowledge about how it is going to be used and what chemicals are involved to determine the safety and risks.

Here are some of the possibilities that have been discussed over the past several years:

Predictive containment is a good place to start. As discussed in Chapter 6, face velocity alone is not a good indication of containment because it doesn't recognize the air pressure changes in the laboratory.

Weather forecasting is a good analogy to the complexity of predicting fume hood containment. In looking at weather, two aspects are generally considered -- current conditions and the forecast. The current conditions are a real time reading of what is happening, but the forecast is based on many different factors considered together to establish the probability of an event such as rain. If the weather forecast says a 40% chance of rain at 2 p.m., that is based on watching the trends of many factors. This same approach can be applied to fume hoods. A number of factors impact the hood's performance. These factors include things like face velocity at various locations, differential pressure between the inside of the hood and the room, sash position, temperature gradient, exhaust volume, static pressure, slot velocity (velocity in the plenum behind the baffles), as well as other factors. While none of these alone is an indication of containment, when considered together they provide a good basis for forecasting containment.

Predictive containment would employ many sensors in and around a fume hood to provide a large data stream with a robust overview of the current conditions. Just like the weather forecast, this data could be modeled and a forecast of the likelihood of containment could be made. This real time monitoring could be used to alert the user. By using a simple green, yellow, red LED strip, anyone in sight of the hood could be alerted to the probability of containment loss. If the probability of containment is higher than say 70%, a green light could be displayed. If the probably drops to 40% to 69% a yellow display could be used and if the probability drops to 39%, a red display would notify users of the danger. This type of modeling would provide those in the laboratory with a good real-time indicator of fume hood performance. As the probability of containment goes down and the user is alerted, there is an opportunity for the user to evaluate the situation and see if corrective actions can be taken. Remember that 25% of loss of containment issues are driven by the user. The indication does not have to be limited to a local display, it could be viewed on a smartphone application or notifications could be sent to various monitoring devices or personnel.

This data stream can be recorded for further analysis. For example, if a group of fume hoods showed a high frequency of yellow and red conditions, it could be indicative of a system problem. On the other hand, a single hood showing frequent yellow and red conditions, might suggest a usage problem or a single hood malfunction.

While this concept does not provide the ability for a fume hood to self-correct, it does alert the stakeholders to potential problems. This could also provide a platform for future fume hood enhancements, but for now, predictive containment could be adapted to both new and existing hoods. It could be a good upgrade for exiting velocity alarms to give users real-time visualization of performance.

The future is not a destination, it is a journey. An interesting technology journey to study is the smartphone. In 2007 when the first Apple phone was introduced, there was no idea where the technology would take us. Now today we are totally reliant on it. While predictive containment is an important goal, it is only the first step on the journey towards the "smart" fume hood.

I have been fortunate to be able to observe technologies that are currently being worked on and am personally excited about the future and the benefits I see for laboratory users, but since much of what I have seen or worked on is proprietary, I can only discuss it in general terms. Suffice it to say that there are many developments on the horizon.

Numerous patents have been issued and much research and development are being performed. A more holistic approach is being taken. We are looking for new solutions to problems that have existed for decades. Central to the solutions is the fact that a fume hood is only a component in a bigger more complex system. To address this challenge, you must either look at the entire mechanical system, uncouple the fume hood from the system or both. Historically, fume hood designers were only able to address performance issues by attempting to design a more robust fume hood that would perform effectively under a wide range of conditions. Today however, fume hood designers have many more tools at their disposal and many of these designers have the vision to take laboratory ventilation to a much higher level. Even as these new fume hood technologies are being developed, much can be done to make current technology more robust.

I can't say it enough, user education and training is as important to a good outcome as the actual technology. Just as now, for new technologies to be successful, the user must be engaged and educated to be a major participant in the ultimate solution.

One of the biggest challenges with traditional hood design is the vortex created in the fume chamber by the baffles and or the sash. These vortices have a tendency to sweep up contaminants and move them closer to the users breathing zone. Minimizing the vortex, controlling it or otherwise eliminating it, greatly improves the fume hood's ability to contain. There are several concepts that work to control, minimize, or eliminate the vortex. These technologies all work differently. Some designs create multiple zones within the fume hood that support the vortex and keep it more at the top of the chamber. Other designs create alternate pathways for airflow at the top of the fume hood that eliminate the vortex. The intent of these new designs is to minimize turbulence, particularly at the worksurface.

There is currently a product on the market that uses an inclined air curtain as the barrier to control containment. There are no baffles in this fume hood design, instead, there is a slot at the top rear running the full width of the hood. What is unique about this design is that unlike most fume hoods that have a 3-dimentional airflow, this design is 2-dimentional. As a result, this fume hood design can be sized to various widths with little effect on performance, unlike a traditional fume hood design where each width has a very different performance profile.

While these technologies create a more robust hood, they don't address the root cause of most problems, which is the management of the room pressure. As long as a fume hood is totally dependent on supply air from the room for its makeup air, there will be issues related to the time it takes to balance air pressure - resulting in the potential for loss of containment during these readjustment periods. By making the fume hood "smarter" and enabling it to communicate directly with the room air controller, the hood becomes more of a working partner in the system. This is one way to improve performance, but will not resolve all the issues. Smart fume hoods can talk to each other and to the room controller creating a conversation and allowing more interaction.

While bidirectional communication is good, for a fume hood to be independent and truly standalone, it will need to have control over both the exhaust and the supply air. This concept is called synchronized-supply. In this technology, a fume hood has control over the supply air and the exhaust and is able to regulate them in a way that maintains an ideal differential pressure within the fume hood. While the early versions of this technology were a step in the right direction, those implementations lacked the intelligence to truly optimize the hood's performance.

What makes a fume hood "smart"? What is intelligence in a fume hood? A smart fume hood starts with the device becoming self-aware. The fume hood needs to be able to detect what is happening in and around it. Basic intelligence starts with the addition of a sensor array that collects data on the current conditions. The next level of intelligence is to process that data and determine what is actually happening in and around the hood. Is the fume hood being used or is it sitting idle? Is the sash open or closed? Are people in front of the hood working? What is the status of the laboratory, occupied or unoccupied? Are people coming and going and opening and closing doors? The more information the fume hood is able to collect and analyze, the "smarter" the hood will be. As it becomes smarter, the fume hood can use that information to better control its performance.

In the drone world, a group of drones is referred to as a flock or a hive. These groups work together to collect and develop information. The same principle can apply with a group of hoods in a single laboratory space working together. Collectively, a group or hive of fume hoods can know more about the surroundings than a single fume hood. One room of hoods can communicate with another room of hoods and create a network that communicate with the Building Automation System (BAS).

Let's consider how a fume hood is used today and how that experience could change with the fume hood of the future.

Today in a typical fume hood use scenario, we have a person using the hood. There is the physical setup inside the hood along with the chemical process that is being performed. The user has total control over the sash position. They can choose to wear protective equipment or not. Much of what happens is at the user's discretion and is affected by their discipline at following procedures. Oftentimes there are notations on the sash of the fume hood describing what is being done in the hood, who is doing the work, and the user's contact information. Again, this is an optional, not a requirement. While much of this is intended to make the experience as safe as possible, there is little actual feedback on the hood's performance.

How many times have we gone into a laboratory only to find the hood alarms muted because the user can't stop it from alarming and doesn't want to keep hearing the alarm sound? How many unsafe setups have we observed? I have seen as many misuses of the fume hood as I have good uses. A more intelligent interaction between the fume hood and its user would be helpful.

The next generation of smart hoods will likely be called cobotic hoods. This term is a combination of the words, collaborative and robotic. For years, industrial robots did their jobs with no awareness of what was happening around them. It was necessary to fence off the robots so that people and materials could not get in their path of operation. The cobot is a robot with self-awareness. It is able to sense its surroundings and adjust its behavior to fit its programing. It can avoid people or objects in its path.

A cobotic fume hood would possess the self-awareness necessary to provide operational data to the backend systems, so that a host of variables could be analyzed and performance of the entire space optimized. By humanizing the interface, the technology is more accepted by the user and the cobot is more likely to be considered a friend rather than a foe. A cobotic hood can become a virtual/digital laboratory assistant that helps the user be a responsible and more predictable component of the overall system. This new technology will have the ability to minimize danger and allow the user more time to focus on the science and less time on the routine tasks.
Today, there are a few laboratories where cobots work side by side with laboratory technicians. They perform many of the repetitive and boring tasks. At some point, these cobots will take over hazardous work. They will interface with the fume hood. To accomplish this there must be a method for the cobot to communicate with the fume hood.

Another automation coming to the laboratory is telepresence robots. These are remotely controlled, wheeled devices that use wireless Internet connectivity to communicate. Laboratory collaboration using telepresence is expected to become common. It also represents an opportunity for you to be in your lab without physically being there. Cobots are already using instruments and equipment and as more of these devices appear in the laboratory, there will need to be more communication between these digital devices.

The future fume hood will have more of these digital features. In fact, it is likely that the term fume hood will be abandoned for a more accurate term. Perhaps a more descriptive term such as chemical containment device (CCD) will be adopted. Some manufacturers are already using this term. The CCD should assist and help the user make better decisions on how to use the device. While a CCD could assist multiple users, the interaction should be tailored to each person and what that person is doing at the time. As an aware device the CCD should consider what other activity is taking place in the laboratory. The performance variables of the fume hood can be adjusted based on risk.

What might an interaction with a cobotic fume hood look like?

It all starts with a new interface. The concept is to create a digital partner that can assist the user in making good choices. The interface of a CCD will most likely be a tablet like interface that can be both a touchscreen that responds to a gloved hand and hand movements as well as voice commands. We want to humanize the interface as much as possible.

The CCD can provide a new level of security and control, making the experience safer and more productive. The first principle is access control. Access is controlled by a combination of radio-frequency identification (RFID) and/or Face/Voice recognition. So, the CCD recognizes the user and based on this recognition, the CCD can access information about the user such as training records. Knowing who is using the device is important to be able to personalize the experience. This data will also be useful in the back office to better understand usage.

In our intelligent laboratory, we want to humanize technology. The goal is for the devices to not only protect the individual, but to assist the person in making good choices and being more productive. The control system would have almost unlimited access to data.

The interface could be similar to Amazon's "Alexa" or Google Assistant's "Ok Google." To allow many people to work together in the same area, a unique action phrase or ID for each device would be used. In this example we will call our fume hood/CCD "Number 6."

The same technology could be expanded to other systems around the laboratory, much like the home automation hubs. The interface could be personalized to make the technology feel more comfortable or human.

"A future Smart Laboratory Interaction"

Assume that your name is Mark and you have named your digital assistant Abby. Abby is available on any intelligent device in the laboratory, in fact it could be accessible through any Bluetooth device.

You approach the locked door to the laboratory. The entry system is equipped with a thermal scanner and facial recognition. As you approach, the voice you have chosen for Abby says "Good Morning Mark." Or you could change your preference to provide you with a more personal or expanded greeting such as, "Welcome back Mark, remember you have a staff meeting at 10 a.m. this morning."

You respond, "Good Morning Abby" as you enter the laboratory. You proceed to the area where you work (your office, your work bench, your write-up area, or a specific bench area that has been assigned for the day), and when you are detected in that area, Abby would say, "Mark, what can I help you with?"

You might say: "What is on my schedule today?" or "What is the weather?" or "Is Joe Smith here yet?" or "Set reminder for ..." or "Schedule a meeting for 10 am with Joe Smith" or "Is Number 6 available?" This is a similar type of conversation that you might have with smart devices such as Amazon Alexa or Google Assistant.

Next you go into the laboratory and approach a fume hood (now called a CCD) where you intend to work. This CCD has been named "Number 6" (Each device has an individual name so several CCDs or smart devices can co-exist in the same space).

When you approach, Number 6 says, "Good Morning Mark, what are we doing today?"

You respond with your work plan.

Number 6 says, "Mark, in checking your files, I see that you haven't received current training for that process, can I show you a short training video?"

You respond, "Sure."

After the video training is complete, Number 6 asks you a few test questions to verify your understanding of the training material. You answer those questions. Then Number 6 says, "Mark, do you acknowledge that you understand the procedure and the risks?"

You say, "Yes I understand."

Number 6 says, "Ok, I will record in your file that you now have received the training."

Number 6 says, "This work requires the following personal protective equipment (PPE): latex gloves, safety goggles, and your laboratory coat. Please confirm you are using these items."

You say, "Yes I have my gloves and goggles and laboratory coat on."

Number 6 says, "I have calculated that the risk level for this procedure is a 4. I am adjusting the sash and airflow parameters to ensure your safety. Are you ready for setup?"

You say, "yes."

The sash opens and the lighting adjusts to a setup level. The CCD now shifts from unoccupied mode to setup mode.

A laser projects information on the worksurface showing the safe work areas. A setup checklist is presented on the display.

Number 6 asks, "Do you need assistance with the setup?"

You say, "Show me a typical setup for this task."

A photograph or diagram is shown on the display and a laser guideline is projected on the work surface. If this procedure/experiment has been done before, a photograph and other related information can be stored and displayed again. Photographs or reference materials can be called up by the user as needed, along with the Safety Data Sheet (SDS). At this point, the system knows who is using the CCD, what task will be performed and the conditions in and around the CCD. The internal camera takes photographs and video records the activity. The video loops every hour and only saves the footage based on preset conditions such as an accident or a red condition lasting more than a specified period of time. A photograph is taken every time the CCD changes mode and another each hour.
You perform the appropriate setup.

You say, "Ok Number 6, I am finished with my setup and am ready to start."

Number 6 says, "Would you like to see an airflow visualization?"

You say "Yes."

The sash adjusts to the operating position and the CCD switches to operational mode. The test fog starts coming out of test ports and you can see the impact of this setup on the airflow. After a minute, Number 6 asks, "Are you satisfied with the setup?"

You say "Yes." The fog stops and you are ready to start.

Number 6 asks, "Would you like me to send you notifications about this experiment as it progresses?"

You say, "Yes."

You go back to your office and monitor the experiment via video receiving real-time data and updates on the progress of your project.

While you are away, the monitor on the fume hood displays what procedure is being performed along with your contact information. If the CCD was outfitted with a cobotic arm, some of the steps normally performed by a human could be automated.

The CCD knows what is happening on the inside of the fume chamber as well as within the laboratory room in general. An exposure risk factor has been calculated and CCD performance has adjusted accordingly. The CCD is logging data from all sensors. Because many laboratories will have multiple CCDs operating at the same time, the combined data is shared in an Internet of Things (IoT) style, allowing more interaction of devices and better predictions.

At the CCD, the display is a graphic that shows what is happening inside the hood and the background is either, green, yellow, or red based on the predictive containment calculations -- Green indicating good containment, yellow indicating caution, or red indicating the danger of containment loss. The display shows the user's name and contact information. It also has an emergency button. It might additionally display elapsed time, or time remaining, or other critical data.

This display is visible to others in the laboratory so that they can see who is using the CCD, what is being performed, and the predicted containment level along with other useful information.
Assuming that out of a 7-hour day, the CCD is not used for 2 hours, and is in setup mode for 1 hour and only working 4 hours, the CCD can automatically adjust from standby mode, to setup mode, to operating mode for a level 4 risk. In those 7 hours the CCD might have used 50% less energy than used today in a typical system and the

laboratory will be safer. Because the CCD is communicating with the other CCDs and the users, it can anticipate system changes, minimizing system response time and greatly improving the probability that the CCD will remain in a green condition.

The collection of sensor data and user inputs is a large amount of data that is sent to a data lake for storage. The amount of data collected makes the CCD a "Big Data" device. Using Big Data techniques, the data can be mined and reported in an easy to use dashboard.

Examples of situations in which this data could be useful include:

A particular CCD has been showing a high number of abnormal red conditions. The data shows that a particular user is present for most of the events. By reviewing the setup photographs and other related data it can be determined whether this problem is related to a training issue that can be addressed with the user or whether it is a malfunction of equipment.

A particular laboratory room is showing a high number of red conditions from multiple CCDs. Analysis of the data might suggest a balance problem or equipment issues that can be addressed by maintenance.

This data can provide a more holistic view of the laboratory and its operation. The data can be trended to show user patterns and requirements for services. With this data platform, the laboratory can be operated at a higher level of safety, reducing risks, reducing energy usage, and generally increasing efficiency and productivity.

This is just a quick overview of how a smart CCD might behave and interact with a person. The introduction of artificial intelligence (AI) and machine learning will change the way we think about laboratory space and how we manage it.

While not all the functions will be present in the first generation of smart fume hoods, they are all realistic possibilities as the technology in laboratories evolves. While it may take several generations of products to advance to this level, we can expect continual progress. Much like smartphones where we see a new generation every year, each new model of CCD will have more features and capabilities.

It will probably take a decade for the concept to mature, but if you can see a future, you can make it happen. Our goal is identifying the problems we are addressing and then look for innovative solutions. The smart hood will be like many other smart devices. Different manufacturers will introduce different features in different generations of the product. Like smart phones and smart devices, designers of smart CCDs will learn from the market what features are most effective and desired. Innovation and progress will build on the successes and each generation of product will improve.

Again, for the fume hood/CCD to function safely, it needs to know what is happening in and around it as well as how the laboratory ventilation system and the various building systems are performing. Instead of being just a component in the mechanical system, it will become smart and an active participant in the safe operation of the laboratory.

Over the next decade these are some of the features we could see in fume hoods of the future:

- These intelligent CCD devices will be outfitted with various sensors to monitor various conditions and predict performance. There is nothing intelligent about today's fume hoods, they are just boxes that exhaust. Much like a robot, they are totally unaware of anything around them. They can perform as programmed, but the performance is mostly static and has little to do with the risks or with what the fume hood is actually being used for. Thus, we are dependent on the user to be smart and use the hood safely. But even a safe user has minimal control over the system performance.

- The concept of a smart CCD is to be self-aware much like a cobot. The CCD needs to be able to sense what is going on inside the hood and around it so that the risk of exposure to the users can be assessed. The ability to communicate with the laboratory ventilation system is important so that the CCD can manage the operational parameters to maximize safety and minimize energy usage.

- As an IoT (Internet of Things) device, the CCD will have the ability to provide notifications, alarms, texts, and email, but more importantly, it will be able to communicate with other fume hoods and laboratory automation systems and building systems.

- To be able to verify system performance and containment, the CCD will monitor the indoor air quality with a built in volatile organic compounds (VOC) monitor.

- Taking the lead from insurance company use of dashcams in cars, the data gathered by CCDs will be used to minimize risk. Video in front of the CCD and video inside the CCD will provide a record of activity. Just like dashcams and security cameras, events can be tagged for possible review. The resulting data can be used for training and risk reduction. Similar to flight recorders in airplanes, the data can be used to determine what went wrong when a malfunction is detected.

- Outside the CCD, an occupancy sensor will be able to determine if someone is standing in front of the CCD. If so who? Are others near the CCD? What are traffic patterns around the CCD? Part of being self-aware is understanding the surroundings so that performance can be optimized.

- Radio-frequency identification (RFID) or biometrics will allow the device to identify the person who wants to use it or is using it. The CCD can check training records to see what the user's training level is. It will be able to record the time they spend operating the CCD and what chemicals they are working with. This data might prove valuable in the case of an accident or illness.

- The ability to change LED lighting color from 2800K to 6500K and intensity from 0 to 100 percent makes it possible to light the CCD in a way that is comfortable and productive.

- Motion detection within CCD can alert the user to possible events such as a piece of equipment falling over. This will also detect the degree of user hand movement within the CCD and can respond accordingly.

- A digital interface for controlling electricity, laboratory gases, and water to improve safety in the laboratory. The CCD will need to be able to control these services. As Cobot workers become more common in laboratories and work successfully alongside people, they will need to interface with the CCD and control the functions like the sash, the electricity and the laboratory gases. This will be done wirelessly using digital commands.

- The use of multiple RFID sensors can reveal the exact location of people in the laboratory. Knowing people patterns can help us better manage the laboratory. The traffic patterns and heat generation can be better predicted. Laboratory occupancy could be a factor in controlling air changes and temperature. Today collecting this information is not feasible, but with the CCD sensors, this now becomes realistic.

- Synthetic Sensors are becoming more common and could become a component in a smart CCD. If you want know about this technology and what it can do, take a look at this short YouTube video (**https://youtu.be/aqbKrrru2co**). This video shows you how simple general-purpose sensors can provide extensive information about the environment they are placed in.

- Knowing the differential pressure changes from area to area can reveal much about the air movement within the building itself. We know that a temperature gradient can impact the pressure gradient. Switching from heating to cooling can have an impact on performance. The more data collected, the more accurate the view of what is happening. This will allow the variables to be controlled. A more robust monitoring of DP will enhance performance and safety

- Knowing sash position/configuration is critical for several reasons. With the CCD controller largely controlling the position, the impact of sash position will be less of a factor. Combination sashes (both vertical and horizontal) need to have the same level of control as vertical only sashes. The most likely system for accomplishing this would include the horizontal panels having small servo motors to close the horizontal panels when the vertical sash is opened. An alternative to this would be to have a dual sash system where the horizontal panels are located inside the CCD behind the vertical sash. Another alternative is to have safety panels mounted exterior to the sash that can be moved into position to function much the same way as combination panels are used today.

- Average face velocity is still important, but in a future CCD system this is not very actionable data. Given the lack of training in most laboratories, it is unlikely that most users will know the significance of the face velocity number in predicting containment. Controllers will be more graphic and will look more like tablets. The graphic user interface will simplify the communication between the user and the CCD. Instead of a face velocity alarm the user will see much more useful and actionable information.

- While not specifically useful to the user, knowing much more about what is happening within the laboratory ventilation system will help optimize CCD performance. Today much of this data comes from sensors that are only calibrated occasionally and the system adjusted periodically. With more sensors and real time data, the overall laboratory environment can more readily be adjusted and optimized. Everything from temperature, lighting, humidity, noise and air quality can be monitored and adjusted to provide more user comfort along with increased safety and reduced energy consumption.

- To optimize containment, the CCD needs to know about cross drafts at the face of the fume hood. These disrupt laminar airflow into the fume hood, so detecting cross drafts is important to predicting containment. Additionally, other sensors are needed to detect turbulence. Turbulence detection is another important feature of the sensor array.

- To better understand how the laboratory ventilation system is performing in real-time, additional system information is needed. Monitoring the static pressure at the point where the CCD connects to ductwork not only reveals the load on the system, but changes in static pressure also indicate what is happening in and around the fume hood. In isolation this data may have little use, but when considered in context with other data it gives us insight to overall system performance. Additional useful information would include the duct velocity. Duct velocity not only has an impact on duct noise, but it also must maintain a minimum transport velocity to keep the contaminants in the air stream. In addition, knowing the total static pressure at the blower tells us the percentage of capacity we are using and is an indication of performance as well as energy usage. Again, coupled with other data this is another piece of the puzzle. The more we know, the better we can understand problems and find solutions.

- When we are considering re-entrainment and downwind impact of the fume plume, knowing the exhaust velocity is very important. As we become more aware and concerned about what is being discharged into the atmosphere, we will want to know a lot more about the content of the discharge and where it is going. Conditions on the roof such as temperature, wind speed and direction will help us take control over the discharge stream.

- Knowing the type of work that is being performed in the CCD is important in understanding the risk factor. This can be user entered data or a combination of sensor data and user data. The purpose of this data is to determine the risk factor for exposure and danger level from exposure. In a smart system, the CCD could adjust performance to match the risk level.

- The user's training level can be obtained from the laboratory records and it shows whether the user is trained and qualified to use the CCD and whether the user has training specific to the procedure to be performed. The smart CCD can offer point of use training and safety reminders.

- Differential pressure between the inside of the fume hood and the laboratory is one of the best indications of momentary loss of containment. Strong negative pressure within the fume hood is a good indicator of containment when factored with other variables

- Room Temperature and humidity impact the air movement. Temperature gradients within the room impact turbulence and cross drafts. Monitoring and understanding these factors can help improve CCD performance.

- The temperature inside a fume hood is a micro version of what is happening in the laboratory. Monitoring this temperature helps predict turbulence and potential problems associated with the process, such as excessive heat or fire.

- Automatic sash movement will eliminate a safety risk. The sash not only becomes a physical safety barrier, it becomes a means of access control. Remembering to lower the sash to the operating position or to close it when finished becomes a thing of the past.

- It will be possible for the CCD to maintain a usage log automatically and to summarize any level of usage detail. Information such as who used the CCD, for how long and how many red conditions occurred during their usage will be gathered for later analysis.

- The ability to generate fog will allow the visualization of airflow within the hood and show the impact of the equipment setup on airflow. This tool allows the user to make adjustments to the setup before loss of containment occurs.

- Laser projection of images on the worktop, from simple boundaries to complex setups, can assist the user in fast and proper setup.

- Because the CCD is voice controlled, users will be able to use voice controls to record information and data. Voice controls can also be used to provide information, record notes and log data.

- Future CCDs will interface with building systems (general HVAC, laboratory ventilation, fire, security).

- The ability to integrate augmented reality (AR) and virtual reality (VR) into the CCD setup and procedures brings about a whole new experience.

These are just some of the features that can be performed by a smart CCD. Until these smart units are put into use, we will not know what features are most desired and effective, but the trend to use smart devices has already started. It is clear that the fume hood as we know it today is problematic at best and as the technology improves innovators and designers will use new technology to go outside the box to find better solutions to the problems. For the first time in a hundred years, the next generation of smart devices are within reach. Finally, fume hoods are poised to live up to their intended use of providing maximum safety to users while being environmentally sensitive and minimizing energy usage.

The year 2020 and COVID-19 have brought about many changes in how we view and use technology and how we work. The digital transformation has advanced at least 5 years in a single year. The impact of how we will work during a pandemic will impact laboratories for many years to come. As a group, I think we will be more open to technology as a way to more effectively work in the future. The items discussed above are not science fiction, many are being developed today or are already in the market. The future is now.

Chapter 15

The History of Fume Hoods

While working at Kewaunee Scientific during the 1980s, I was given a project to research the company's history and the history of the fume hood industry. I found the project intriguing for a number of reasons. First, it was a very interesting chain of events that led to the current status of the fume hood market. Equally enlightening was the realization of how much can be learned from the past. The reasons behind current fume hood practices and the analysis of past failures and successes provide valuable lessons. To preserve this legacy, I am ending this book with the history of fume hoods. I hope you find it as interesting as I do.

Timeline for the development of fume hoods.
The need for ventilation has been apparent since the early days of chemical research. The following is a chronology of important milestones in fume hood development.

600 AD to 1600 AD -- The fume hoods of this period were fireplaces used by alchemist in the middle ages. The origin of the modern fume hood lies in the fireplaces of ancient alchemists who practiced a combination of chemistry, physics, metallurgy, and mysticism, often in attempts to convert base materials into precious metals.

1790 -- Joseph Priestley designed a chemical exhaust system for his laboratory in Pennsylvania operated by large manpowered bellows.

Early 1900s -- One of the earliest scientists concerned with laboratory ventilation was Thomas Edison who used the fireplace chimney to exhaust noxious fumes and odors from his laboratory. Thomas Edison used a fireplace as a fume hood on cold winter days, but on warmer days Edison would place a shelf outside his laboratory's double hung vertical window. This was the genesis of today's fume hoods.

1923 -- The first recognizable fume hood, in the modern sense of the word, was developed at the University of Leeds. It consisted of a large cabinet standing at working height and incorporated vertical rising sashes arranged like parallel windows. It was a takeoff on Edison's fume hood and was designed by interior decorators as a cupboard using vertical counter weighted sash widows. This device was referred to as a fume cupboard. To this day, in Europe, fume hoods are called fume cupboards. This laboratory style furniture is why many consider the fume hood a furniture item rather than the mechanical device it is.

1939 - 1945 -- During World War II considerable advances were made to fume hood technology in response to fears of toxic chemical exposure and radiation. This resulted in improvements in design, safety and ventilation. The first real fume hood design changes occurred due to the numerous deaths of researchers working on the atomic bomb project. Designers learned anecdotally that no particular face velocity made a fume hood safe, but they never understood why. It took another 50 years to learn the reason why fume hood face velocity alone did not make a laboratory safe.

Also, during the war years, the high-efficiency particulate air (HEPA) filter was developed by a group which later became the Atomic Energy Commission. The development of the HEPA filter had a dramatic impact on the effectiveness of fume hoods and biological safety cabinets, greatly increasing protection for users.

1943 – John Weber, Jr. working at the Ames Laboratory in Ames, Iowa, developed the concept of a constant face velocity, variable exhaust flow fume hood control. This design was applied to a vertical rising sash hood served by a dedicated hood exhaust fan. The concept eventually became a standard feature employed on many fume hoods. At that time, it was used in atomic laboratories, especially where ventilation containment within the hood was critical.

1951 – H.W. Alyea, chief field engineer at the Johnson Service Company (now Johnson Controls, Inc.), realized that keeping the door of a fume hood closed as much as possible, and certainly when not in use, resulted in a considerable decrease in the amount of airflow through the laboratory. Along with this reduction in the air supply air came proportional reduction in cooling demand and considerable energy savings

Early 1950s -- John Turner, working in the Engineering Department at Oak Ridge National Laboratory (ORNL), suggested replacing vertical rising sashes with horizontal sliding sashes in order to reduce energy consumption. He also introduced the use of a mechanical damper that moved as a result of the imbalance between external and internal hood pressures.

Around the same time, Weber also recognized that the best containment in a fume hood was achieved with a minimum sash opening. This awareness resulted as he developed the emergency quick close feature incorporated as a part of his system. The exhaust fans used as a part of Weber's system were never turned off.

By 1960, fume hoods had been established as box-like enclosures with a movable sash. The baffle systems resembled a traditional fireplace and the laboratories they were installed in had pressurized air supply. The exhaust system was never turned off.

The 1960s -- Many suppliers joined the market with different concepts and many products that didn't really work that well. The lack of product standards allowed a wide range of products to come to market. 1961 – Labconco introduced its first one-piece molded fiberglass-lined fume hood. Fiberglass was chosen to line the fume hood because of its durability, cleanability, high light reflectivity, fire resistance and chemical resistance properties.

1964 – Kewaunee produced the Aristocrat Auxiliary Air Fume hood. Kewaunee's work with the government on the Manhattan Project made them a leader in fume hood development. The introduction of auxiliary air fume hoods conserved energy by introducing outside air in front of the hood. This reduced the loss of tempered air from the laboratory. This type of fume hood required the use of two duct and blower systems.

1968 – François-Pierre Hauville created the company Erlab and began selling the first ductless fume hood the same year.

1970s -- Fume hoods constructed from wood were replaced with epoxy painted steel. During this time, the walk-in hood was introduced, as well as a number of special purpose hoods such as perchloric acid and radioisotope hoods.

1980s -- The industry responds to a market flooded with poor products. The writing of fume hood standards began and ASHRAE 110 -1985 was released for testing hoods. As ASHRAE 110 was being specified, companies were required to either improve the performance of their products or exit the market. The end result was that many of the unsafe products left the market. During this time, the precursor to SEFA (SAMA) wrote a fume hood standard.

More changes in fume hood design and performance occurred in the 1980s than in the previous 50 years. Most of the lessons learned were by trial and error, but a full understanding of the laboratory ventilation system and the factors that impact a fume hood's performance were still elusive. Another major change that occurred during this time was the elimination of liner materials that contained asbestos. During this same period there was a huge improvement in testing equipment that made it possible to do containment testing.

1985 -- AFNOR introduces the NFX 15-211 ductless filtering fume hood performance safety standard.
Variable Air Volume (VAV) technology was being applied to fume hoods to conserve energy.

1990s -- Several major improvements to fume hood technology occurred. New material technologies, and requirements for better chemical and flame resistance lead to a number of new design concepts. Flush sills were introduced as well as auto sashes and chain driven sash systems.

1991 – The Occupational Safety and Health Administration (OSHA) was given jurisdiction over laboratories, which had been previously excluded. This government standard OSHA 29CR-1910.1450 was the first government standard that addressed fume hoods.

1993 – Tom Saunders wrote the book "Laboratory Fume Hoods – A User's Manual" in an effort to explain to users how the fume hood was intended to be used.

1994 – SEFA 1 for fume hoods was written. This was an industry standard that updated the previous SAMA fume hood standard. Also, the CSA Z316.5 ductless fume hood safety standard was written.

1995 – A new version of ASHRAE 110 was released that greatly improved the quality of fume hood containment testing.

1996 – The axle, chain and sprocket sash system was developed by Chip Albright.

1998 – Computational Fluid Dynamics (CFD) modeling began to be applied to fume hoods.

1999- Underwriter's Laboratory released UL1805 which provided requirements for laboratory hoods and cabinets addressing many of the safety issues that were being recognized in the market. Additionally, a greatly expanded SEFA 1 Fume Hood Standard was written.

During the 1990s, variable air volume (VAV) hoods began to outnumber constant air velocity (CAV) hoods, especially in laboratories that were 24/7 systems.

2000s -- Driven by demand for more energy efficient models, low flow fume hoods, operating at face velocities of 40-80 fpm, were developed. These fume hoods delivered excellent performance while saving energy and money. However, the concept of "safety first" was overshadowed by concepts that saved energy and money and didn't always provide optimal safety.

2008 -- ERLAB introduced the filtering fume hood technology also known as GreenFumeHood which incorporated Neutrodine multistage general-purpose filtration for use in ductless fume hoods.

2012 – The ANSI Z9.5 standard was updated.

2016 – The ASHRAE 110 standard was updated.

2018 – The concept of synchronized supply brought a new class of fume hoods to the market.

2019 – The EN 14175 standard was updated.

Now we are up to current times, I hope you have learned more about fume hoods and what is necessary for them to provide the intended safety. As I said in the previous chapter, I am very excited about the future of laboratory ventilation and chemical containment devices. Smart hoods are coming and most likely from Asia, but as this history is updated, I see this as a defining moment for fume hoods and laboratory safety.

As I conclude this book, I can honestly say that I am optimist about the future. As more people understand the issues at play and we begin to address core problems rather than symptoms, technology will help us displace many of the bad habits and unsafe practices. User safety will improve and we will be able to make labs greener. Technology will allow us to work smarter and be more productive. Much of the routine and boring tasks can be automated allowing the scientific minds to be more creative and experience greater discovery.

Thank you for taking the time to read this book and I hope you have found it worthwhile. I feel the best way to improve is through education. Hopefully this book will help make you a safer hood user.

Glossary of Terms

These are terms that are used in this book and are common when discussing fume hoods and laboratory ventilation systems:

Acceptable Indoor Air Quality: Air in which there are no known contaminants at harmful levels as determined by appropriate authorities and air which 80% or more of the people find acceptable.

Access Opening: The part of the fume hood through which work is performed; sash or face opening.

ACH, AC/H (air changes per hour), N: The number of times air is theoretically replaced in a space during an hour. An ACH rate for a room can be converted to volumetric airflow by multiplying the ACH number times the gross volume of the room.

Adjacent Roof Line: For the purposes of determining the laboratory chemical hood stack height, the adjacent roof will be within 6 feet horizontally of the nearest exhaust fan stack.

AHU, Air Handler Unit: A device used to regulate and circulate air as part of a heating, ventilating, and air-conditioning (HVAC) system. An air handler is usually a large metal box containing a blower, heating or cooling elements, filter racks or chambers, sound attenuators, and dampers. Air handlers usually connect to a ductwork ventilation system that distributes the conditioned air through the building and returns it to the AHU. Sometimes AHUs provide (supply) and or they can accept (exhaust).

Airflow Monitor: Device installed in a fume hood to monitor the average airflow through the sash opening.

Air Foil: Curved or angular member(s) at the fume hood entrance. The lower airfoil is a horizontal member across the lower part of the sash opening to provide a smooth air flow into the fume chamber across the work surface into the baffles.

Air Lock: An intermediate chamber between two dissimilar spaces with airtight doors or openings to each of the spaces. The doors are interlocked to ensure that at least one of them is always closed.

Air Volume: Quantity of air expressed in cubic feet (ft3) or cubic meters (m3).

Artificial Intelligence (AI) and Machine Learning: Artificial intelligence (AI), machine learning and deep learning are three terms often used interchangeably to describe software that behaves intelligently. This software can evaluate vast amounts of data and make intelligent decisions.

Auto Sash: Is a feature where a fume hood sash has been motorized and can be controlled without using one's hands.

Auxiliary Air: Supply or supplemental air delivered to a laboratory fume hood to reduce room air consumption.

Auxiliary Air Hood: A laboratory chemical hood with an external supply air plenum at the top of the laboratory chemical hood. The auxiliary air plenum provides a makeup airstream comprised of unconditioned or only minimally conditioned outside air to substantially reduce the amount of conditioned room air exhausted by the laboratory hood.

Baffle: A panel located across the rear wall of the fume hood that directs the airflow through the fume chamber. The baffle creates a plenum at the rear of the hood to direct air to the exhaust outlet.

BAS, Building Automation System: Is an intelligent system of both hardware and software, allowing the heating, venting and air conditioning system (HVAC), lighting, security, and other systems to communicate on a single platform.

Bench Hood: A fume hood that is located on a counter height work surface.

Biological Safety Cabinets, BSC: A fume hood like device that is intended for use with biological agents and not for chemical use.

Blower: see Fan.

Blower Exhaust Velocity: The speed at which air is exhausted from the fume hood exhaust stack. The minimum should be at least 3000 fpm (15.2 m/s).

Bypass: Compensating opening in a fume hood that functions to limit the maximum face velocity as the sash is lowered.

Bypass Hood (constant air volume bypass laboratory hood): A laboratory hood design that incorporates an opening (bypass area) in the upper portion of the laboratory hood structure. When the movable sash is fully open, the sash blocks off this bypass area and all of the airflow into the laboratory hood must pass through the open face area. However, as the sash is being closed to reduce the open face area, at a specific point an amount of bypass area is being uncovered. The increase in the bypass area opening offsets the decrease in the face area opening, thus providing an alternate path (the uncovered bypass area) for air to flow into the laboratory hood. When utilized with a constant air volume ventilation system, the bypass area keeps the laboratory hood face velocity relatively constant and from increasing to an objectionably high value as the sash is lowered.

Canopy Hood: Ventilating enclosure suspended above work area to exhaust heat, vapor or odors. This device is not a laboratory fume hood.

Capture Velocity: Speed of air flowing past the face opening through a fume chamber at a speed necessary to capture generated fume vapors and/ or particulates and directed to the exhaust outlet. Measured in feet per minute (fpm) or meter per second (mps).

CCD Chemical Containment Device: A new class of Exposure Control Device that represents the Next Generation of Laboratory Fume Hoods. These devices incorporate new technology not currently present in traditional fume hoods.

CFD (Computational Fluid Dynamics): Is a branch of fluid mechanics that uses numerical analysis and data structures to analyze and solve problems that involve air flows. Computers are used to perform the calculations required to simulate and visualize the free-stream flow of the air within surfaces defined by boundary conditions. With high-speed supercomputers, better solutions can be achieved. A good tool for fume hood design, but generally lacking the definition to truly understand how the air is behaving.

Chemical hygiene officer: An employee who is designated by the employer and who is qualified by training or experience to provide technical guidance in the development and implementation of the provisions of the Chemical Hygiene Plan.

CHP (Chemical Hygiene Plan): Is a written program developed and implemented by the employer which sets forth procedures, equipment, personal protective equipment, and work practices that are capable of protecting employees from the health hazards presented by hazardous chemicals used in a particular laboratory.

Clean Air Hoods: These fume hood like devices are designed to provide product protection. Often found in cleanrooms, they are not intended for user protection against exposure.

Cobotic Hood: This is future generation of fume hood or chemical containment device that is driven by AI and actively participates and assists the user with safe operations.

Color Temperature: Color temperature is conventionally expressed in kelvins, using the symbol K, a unit of measure for absolute temperature. Color temperatures over 5000 K are called "cool colors" (bluish), while lower color temperatures (2700–3000 K) are called "warm colors" (yellowish).

Combination Hood: A fume hood assembly containing a bench hood section and a floor mounted section.

Combination Sash: A fume hood sash with a vertical framed member that moves vertically for setup and houses two or more horizontal sliding transparent viewing panels that can be opened for operation.

Constant air volume (CAV) ventilation system: A ventilation system designed to maintain a constant quantity of airflow within its ductwork. Although relatively simple, a constant volume ventilation system typically requires the maximum ongoing energy usage since the system always operates at maximum capacity.

Counter Top: (See Work surface)

Cross Drafts: Air drafts or currents that flow parallel to or across the face opening of the fume hood.

Damper: Device installed in a duct to control airflow volume.

Data acquisition, DAQ: Is the process of measuring an electrical or physical phenomenon such as voltage, current, temperature, pressure, or sound with a computer. A DAQ system consists of sensors, DAQ measurement hardware, and a computer with programmable software.

Data Lake: Is a centralized repository that allows for storage of structured and unstructured data at any scale without having to first structure the data. Using dashboards for visualization and real-time analytics it supports machine learning and is a guide to better decision making.

Demonstration Hood: A vented enclosure used for student demonstrations that has visibility on at least three sides, used primarily in schools. This device is not a laboratory fume hood.

Design Sash Position: The maximum open area of the hood that achieves the desired face velocity.

Differential Pressure (DP): The difference in pressure between two points of a system, such as between a room and the adjoining hallway, or the fume chamber and the room.

Dilution Ventilation: Ventilation airflow that dilutes contaminant concentrations by mixing with contaminated air, as distinguished from capturing the contaminated air.

Discharge Velocity: The speed of the exhaust air normally expressed in feet per minute (meters/second) at the point of discharge from a laboratory exhaust system. It is the air velocity as it leaves the last element of the exhaust system. The term "stack velocity" is sometimes used when referring to the speed of the exhaust airstream as it is discharged into the outside air.

Distillation Hood: A laboratory fume hood that provides a work surface approximately 18 inches (45.7 cm) above the room floor, to accommodate tall apparatus.

Diversity: Operating a system at less capacity than the sum of peak demand to reduce system size and energy usage.

Diversity Factor: A percentage factor that is applied to establish the theoretical maximum exhaust airflow quantity that is required at any point in time. For example, in an exhaust system consisting of six hoods, it could be assumed that no more than three hoods would be operated at one time. A diversity factor of 50% could be applied. Applying a diversity factor to the theoretical maximum required capacity of an HVAC system is often considered in the design of a VAV system. Incorporating a diversity factor enables downsizing HVAC system components and thus results in a smaller capacity ventilation system. The overall intention of applying a diversity factor when designing a VAV ventilation system is to achieve a lower life cycle cost (e.g., lower system first cost and/or lower system energy costs).

Dual Entry Hood: A bench type fume hood that has two sash openings, usually on opposite sides. These are often used in education and are not a laboratory fume hood.

Duct: Round, square or rectangular tube used to enclose moving air.

Duct Collar: Is the point where the fume hood connects to the duct.

Duct Transition: Is a section of duct that connects two dissimilar shapes or sizes of duct. For example, a rectangular duct to a round duct.

Duct Velocity: Speed or air moving in a duct, usually expressed in feet per minute (fpm) or meters per seconds (mps).

Ductless Hood: A fume hood that is not connected to a laboratory ventilation system, rather, a ductless hood incorporates an exhaust fan and exhaust filters as an integral part of the design and discharges the exhaust directly back into the room. Ductless laboratory hoods are of limited size and capacity in comparison to conventional ducted laboratory hoods. These are for special applications and care must be taken when choosing the correct filter.

ECD, Exposure Control Device: This is any class of product that is designed to minimize the exposure of users to hazardous chemicals.

EHS, Environmental Health and Safety: This is an organizational function whose responsibility is to inspect and monitor environment, machineries and processes to ensure safety as per government rules and regulations and industry standards. They have knowledge of government rules and regulations and prepare standard operating procedures that helps users be safe and healthy.

Electrical Service Fixture: Outlet or other electrical device mounted directly to the face of the fume hood.

Exhaust Air: Air that is removed from an enclosed space and discharged to atmosphere.

Exhaust Collar: Connection between duct and fume hood through which all exhaust air passes.

Exhaust Plume: This is the stream of contaminated air being discharged for the exhaust stack.

Exhaust Stack: Is the tube connected to the output of the exhaust fan. It should have a minimal height of 10 feet or 3 meters above the roof line.

Exhaust Unit: Air moving device, sometimes called a fan or blower, consisting of a motor, impeller and housing.

Face: Front access or sash opening of a laboratory fume hood.

Face Opening: measured in width and height.

Face Velocity: Average speed of air flowing perpendicular to the face opening and into the fume chamber of the fume hood and expressed in feet per minute (fpm) or meters per second (MPS), measured at the plane of the sash.

Face Velocity Monitor: This is a device mounted on the face of the fume hood and displays average face velocity. It often has an alarm when the velocity drops below the programmed minimum.

Facial Recognition: A biometric technology capable of identifying or verifying a person from a digital image.

Fan: Air moving device, can be called an exhaust unit, or a blower, consisting of a motor, impeller and housing.

Fan Curve: A curve relating pressure vs. volume flow rate of a given fan at a fixed fan speed (rpm). This is necessary to properly size a fan.

Filter: Device to remove particles from air.

Friction Loss: The static pressure loss in a system due to friction between moving air and the duct wall; expressed as inches w. g. 100 feet, or fractions of VP per 100 feet of duct.

Flame Resistance: The ability to withstand flame.

Floor-mounted hood (walk-in hood): A larger-size laboratory hood with sash and/or door arrangement that enables access from the floor to the top of the hood interior. The name unfortunately is a misnomer and although the design and height of these hoods may allow it, users should not walk into any hood that may represent a significant exposure hazard. Walk-in laboratory hoods enable larger equipment and apparatus (e.g., equipment on carts, gas cylinders, etc.) to be more readily put in and set up within the laboratory hood.

Fog machine: A machine that generates theatrical fog for airflow visualization in a fume hood.

Fume Chamber: The interior of the fume hood measured in width, depth, and height.

Fume Cupboard: European term for laboratory fume hood.

Fume Hood Controller: Is a device that monitors different elements of hood performance and function. It can control the sash, lights, interface display and VAV for exhaust and supply.

Fume Removal System or Laboratory Ventilation System: Is engineered to effectively move air and fumes consistently through fume hood, duct and exhaust blower. **Note:** Room air, make-up air, auxiliary air (if used) and pollution-abating devices (if used) are integral parts of a properly functioning system and should be considered when designing a fume removal system.

Fume Removal System: A fume hood exhaust engineered to effectively move air and fumes consistently through fume hood, duct and exhaust unit.

Gauge Pressure: The difference between two absolute pressures, one of which is usually atmospheric pressure; mainly measured in inches water gauge (in. w. g.).

GEX, General Exhaust: General exhaust is used to keep the room in balance. Based on the required ACH, if there is not enough air being exhausted through fume hoods and other exhaust devices, air is exhausted through the general exhaust.

Glove Box: A totally enclosed and controlled environment work area providing a primary barrier to confine and contain hazardous materials within. This device is not a laboratory fume hood.

Grille: A louvered or perforated face over an opening in an HVAC system.

Hazardous chemical: any chemical that is likely to be harmful to human health. Often defined by Government regulations and industry best practices.

Heat Resistance: The ability to withstand heat without deteriorating.

Health Hazard: Any chemical that is classified as posing one of the following hazardous effects: acute toxicity (any route of exposure); skin corrosion or irritation; serious eye damage or eye irritation; respiratory or skin sensitization; germ cell mutagenicity; carcinogenicity; reproductive toxicity; specific target organ toxicity (single or repeated exposure); aspiration hazard or simple asphyxiant.

HEPA: High Efficiency Particulate Air (filter) Common standards require that a HEPA air filter must remove—from the air that passes through—at least 99.95% (European Standard) or 99.97% (ASME, U.S. DOE) of particles whose diameter is equal to 0.3 μm.

High Density Shielding: A barrier made of lead. Often used in Radioisotope Fume Hoods

High Heat Hoods: Heat loads in the fume chamber of a hood impacts both the airflow and containment. Some hoods are designed especially for working with high heat loads such as high temperature acid digestion.

Hood: A device which encloses, captures, or receives emitted contaminants. A hood and a fume hood are not the same thing. Often times a hood is vented enclosure.

Hood Entry Loss: The static pressure loss, stated in inches on a water gauge when air enters a duct through a hood. The majority of the loss is usually associated with a vena contracta formed in the duct.

Hood Static Pressure: The sum of the duct velocity pressure and the hood entry loss; it is the static pressure required to accelerate air at rest outside the hood into the duct at duct velocity.

HVAC: Heating Ventilating and Air Conditioning. Ventilation systems designed primarily for temperature, humidity, odor control, and air quality.

Imbalance: Condition in which ratio of quantities of supply air is greater or lesser than the exhaust air.

Inches of Water (inch w.g.): The pressure exerted by a column of water one inch in height at a defined reference condition such as 39°F or 4°C and the standard acceleration of gravity.

Indoor Air Quality, IAQ): indoor air quality related to temperature, humidity, CO2 and airborne contaminants.

IoT (Internet of Things): Is a system of interrelated computing devices, mechanical and digital machines provided with unique identifiers (UIDs) and the ability to transfer data over a network without requiring human-to-human or human-to-computer interaction. The ability to communicate with other devices.

IPA, Isopropyl Alcohol: Is being evaluated as a new way to do fume hood containment testing. It may replace SF6 as the standard for containment testing. The IPA would be atomized inside the fume hood and its escape measured.

Isolator: A core component to the pharmaceutical industry, critical for a range of processes. These gas-tight enclosures provide a complete barrier to ensure aseptic conditions and containment. They are a class of ECD, exposure control devices.

Laboratory: A building, part of a building, or other place equipped to conduct scientific experiments, tests, investigations, etc., or to manufacture chemicals, medicines, or the like. A place to conduct experimentation, investigation, observation in support of research and discovery. An area in which diverse mechanical services and special ventilation systems are available to control emissions and exposures from chemicals and harmful substances.

Laboratory Air Quality: Unlike normal indoor air quality, laboratory air is prone to have other chemical contaminants. The nature of the work in the laboratory has the potential to release these contaminants into the air. Without proper ventilation, the concentration of chemical contaminants can present health hazards.

Laboratory Fume Hood: A Laboratory Fume Hood shall be made primarily from flame resistant materials including the top, three fixed sides, and a single face opening. Face opening is equipped with a sash and sometimes an additional protective shield. Face opening will have a profiled entry and usually an airfoil designed to sweep and reduce reverse airflows on the lower surface. A Laboratory Fume Hood will be equipped with a baffle and, in most cases, a bypass system designed to control airflow patterns within the hood and manage the even distribution of air at the opening. The bypass system may be partially blocked to accommodate Variable Air Volume (VAV) Systems.

Laboratory Module: Is the key unit in any lab facility. When designed correctly, a lab module fully coordinates all architectural and engineering systems. Most laboratory modules are 10'6" wide, but they vary in depth from 20' to 33', depending on the lab requirements and the cost-effectiveness of the structural system. The 10'6" dimension is based on two rows of casework and equipment (each row 2'6" deep) on each wall, a 5' aisle, and 6" for the wall thickness separating one lab from another

Laboratory Ventilation System, LVS: Air moving systems and equipment which serve laboratories. Includes all the mechanical components including fume hoods, exhaust systems, supply systems and all the necessary controls, valves and sensors.

Laboratory Ventilation Design Levels: This is a risk-based system defined in the ASHRAE Classification of Laboratory Ventilation Design Levels Standard. It is similar to design levels used in biological labs (BSL1, BSL2, BSL3, BLS4) The chemical laboratory classifications are LVDL-0, LVDL-1, LVDL-2, LVDL-3, LVDL-4.

Laminar Flow (Also Streamline Flow) – Airflow in which air molecules travel parallel to all other molecules; flow characterized by the absence of turbulence and a Re number less than 2000.

Laminar Flow/Clean Bench (Horizontal Laminar Flow and Vertical Laminar Flow): A clean bench is a laminar flow work cabinet or similar enclosure that provides filtered air across the work surface to protect against product contamination. This device is not a laboratory fume hood.

Lazy Flow or Reverse Flow: When doing smoke visualization, the smoke can be observed moving slowly without a direction. Even when smoke does not escape from the hood, it could show points of weakness or instability in the hood.

LEL (Lower Explosion Limit): The minimum concentration of a particular combustible gas or vapor necessary to support its combustion in air is defined as the Lower Explosive Limit (LEL) for that gas. Below this level, the mixture is too "lean" to burn. The maximum concentration of a gas or vapor that will burn in air is defined as the Upper Explosive Limit (UEL). Above this level, the mixture is too "rich" to burn.

Liner: The liner defines the fume chamber of a fume hood. Usually includes the sides, back and roof along with exhaust plenum and baffle system of a laboratory fume hood.

Local Exhaust Ventilation: Devices that are used to capture and remove emitted contaminants before dilution into the workplace ambient air can occur. Can also be used to remove heat at the source of generation.

Loss of Containment: A condition where the fume hood fails to capture and contain contaminants allowing them to escape into the room air as fugitive contaminants.

Low Flow Laboratory Fume Hoods: Fume Hood designs that provide a reduction in the required exhaust air volume, when compared to the volume required for the same size fume hood to operate with a face velocity of 100 FPM through a fully opened vertical sash. A low flow hood is one that has had the exhaust volume reduced by operating through a smaller sash opening.

Low Velocity Laboratory Fume Hoods: Fume Hood designs that provide a reduction in the required exhaust air volume, when compared to the volume required for the same size fume hood to operate with a face velocity of 100 FPM. Low Velocity Fume Hoods are also referred to as High Performance Fume Hoods and High Efficiency Fume Hoods.

Make-Up air (replacement air): Any combination of transfer air and air provided by a mechanical ventilation system to replace air being exhausted from a laboratory ventilation device. Air needed to replace or compensate for the air taken from the room by laboratory fume hood(s) and other air exhausting devices.

Manifold: A fitting or pipe with many outlets or connections relatively close together. In a fume hood application this is several hoods on a single exhaust duct.

Manometer: Device used to measure air pressure differential, often a u-shaped glass tube containing water or mercury. There are also digital manometers.

Mechanical System: Is all the components that support the HVAC system, ductwork, exhaust and supply systems and all the controls.

Microorganism: A microscopic organism, usually a bacterium, fungus, or protozoan. All of these are to be handled in a BSC not a chemical fume hood.

Minimum Transport Velocity (MTV): The minimum velocity which will transport particles in a duct with little settling; the MTV varies with air density, particulate loading, and other factors.

Natural Ventilation: The movement of outdoor air into a space through intentionally provided openings, such as windows, doors, or other nonpowered ventilators, or by infiltration. Fume hoods will not function safely with natural ventilation. For safe fume hood operation, mechanical supply is required.

Negative Air Pressure: Air pressure lower than ambient.

Occupancy Sensor: This is a motion sensing sensor. It detects human movement. It can be used to detect a user working at a fume hood.

Occupied Zone: The region within an occupied space between 3" and 72" above the floor.

Odor: A quality of gases, vapors, or particles which stimulates the olfactory organs; typically, unpleasant or objectionable. In a laboratory if you can smell chemicals there are generally unhealthy concentrations of contaminants in the air.

Outdoor Air (OA): "Fresh" air mixed with return air (RA) to dilute contaminants in the supply air (SA).

Pa, Pascal: The unit for measuring vacuum or pressure in the International System or SI. One pascal equals one newton (SI unit of force) per square meter.

Particulate Matter: Small, light-weight particles that will be airborne in low velocity air [approximately 50 fpm (.25 m/s)].

Perchloric Acid Hood: A laboratory hood constructed and specifically intended for use with perchloric acid or other reagents that may form flammable or explosive compounds with organic materials of construction. A perchloric acid hood as well as its exhaust system must be constructed of all inorganic materials and be equipped with a water washdown system that is regularly used to remove all perchloric salts that may precipitate and collect in the laboratory hood and in the exhaust system. The exhaust fan must also be of a spark resistant design to ensure against ignition of any perchlorate deposits in the exhaust system.

Pitot Tube: A device used to measure total and static pressures in an air stream.

Plenum: A chamber used to distribute static pressure throughout its interior.

Plenum Chamber: Chamber used to equalize airflow.

Polyethylene: A plastic polymer of ethylene used chiefly for containers, fittings and sinks.

Polypropylene: Material is a polyolefin which is generally high in chemical resistance. This material is commonly used for acid waste piping as well as for deionized water.

Polyvinyl Chloride (PVC): A water insoluble, thermoplastic resin derived by the polymerization of vinyl chloride used chiefly for containers, fittings and piping.

Polyvinylidene Fluoride (PVDF): Material is a strong and abrasion resistant fluoropolymer. It is chemically resistant to most acids, bases and organic solvents, and is the preferred material for piping and faucets for ultra-pure water. Pure PVDF is an opaque white resin that is resistant to UV radiation, and is superior for non-contaminating applications.

Positive Air Pressure: Air pressure higher than ambient.

Posts (columns, facia, pillars): These are the vertical members of the fume hood face. Generally, they form an angular or radiused entry into the fume hood. They are also used for mounting the electrical and service fittings. They work with the upper and lower airfoils to direct air into the hood in a nonturbulent way.

Predictive Containment: Is a forecasting system that uses sensor output and processes it to predict the likelihood that the fume hood is safe and has no loss of containment.

Pressure Drop: The loss of static pressure between two points; for example, "The pressure drop across an orifice is 2.0 inches w.g."

Radioisotope Hood: Provides protection from applications requiring the use of radiochemicals. It has a fully sealed integral work surface reinforced to support lead shielding and coved interiors to facilitate decontamination.

Recirculation: Air removed or exhausted from a building area and ducted back to an air-handling system where it is mixed with outside fresh air. This air mixture is then conditioned and utilized for ventilation. Since air removed from a space is more often close to the temperature and humidity of the building interior than the outside air, the recirculation process enables achieving a greater reduction in heating and cooling energy than once used air where 100% outside air was utilized.

Reentrainment or reentry: The flow of contaminated air that has been exhausted from a building is drawn back into the building through air intakes or openings in the walls or doors and windows.

Register: A combination grille and damper assembly. These supply air outlets can also be called diffusers or grilles.

Relative Humidity (RH): The ratio of water vapor in air to the amount of water vapor air can hold at saturation. A "RH" of 100% is about 2.5% water vapor in air, by volume.

Remote Control Valves: Valves usually installed in the walls of fume hoods with the control handles normally on the face of the hood which regulate and control the flow of the services to the outlets in the interior of the fume hood.

Replacement Air: (Also, compensating air, make-up air) Air supplied to a space to replace exhausted air and maintain balance.

Respirable Particles: A collective group of fine solid particles, aerosols, mist, smoke, dust, fibers and fumes are called Respirable particulates.

Return Air: Air which is returned from the primary space to the AHU for recirculation.

Reynolds number (Re): Is an important dimensionless quantity in fluid mechanics used to help predict flow patterns in different fluid flow situations. The Reynolds number is used to determine whether a fluid is in laminar or turbulent flow.

RFID, Radio-frequency identification: Uses electromagnetic fields to automatically identify and track tags attached to objects. An RFID tag consists of a tiny radio transponder; a radio receiver and transmitter. When triggered by an electromagnetic interrogation pulse from a nearby RFID reader device, the tag transmits digital data, usually an identifying number, back to the reader.

Room Air: That portion of the exhaust air taken directly from the room.

Room Air Balance: A general term describing the requirement that a laboratory room have the proper relationship with respect to the total exhaust airflow from the room and the supply makeup airflow. The relationship of these airflows also establishes the pressure differential between the laboratory room and adjacent rooms and spaces.

Room Balance: In a laboratory we usually want the room to be under slight negative pressure. That means we balance the room to have slightly less supply air coming into the room in relationship to the air being exhausted.

Room Controller: This is an electronic device that processes data from various inputs and then communicates with the Laboratory Ventilation System to manage the exhaust and supply air to maintain room balance.

Room Ventilation: The volumetric airflow through a room expressed in terms of cfm or L/sec.

Sash: A moveable panel or door set in the access opening of a fume hood to provide access, and to provide a protective shield.

SCFM (Standard Cubic Feet Per Minute): Airflow rate at standard conditions; dry air at 29.92 inches Hg gauge, 70 degrees F.

Scrubber, Fume: A device used to remove contaminants from fume hood exhaust, normally utilizing water.

SDS, Safety Data Sheet: Previously called a MSDS sheet. A safety data sheet is a standardized document that contains occupational safety and health data. The International Hazard Communication Standard (HCS) mandates that chemical manufacturers must communicate a chemical's hazard information to chemical handlers by providing a Safety Data Sheet. SDS's typically contain chemical properties, health and environmental hazards, protective measures, as well as safety precautions for storing, handling, and transporting chemicals.

Self-aware: Is an awareness of self and your relationship to the external world. A self-aware device is not only aware of what it is doing, but is also aware of its surroundings. The programming not only looks at what the device is doing, but also considers what is happening around it.

Sensor Array: Is a group of sensors that are collectively taking multiple readings for a complex view of a situation.

Service Fixture: Item of laboratory plumbing mounted on or fastened to a laboratory fume hood

Setback: Is a setpoint below normal. For example, a system might have a night setback. When it goes into the setback mode, it uses a different set of parameters than it does during day mode.

Shall: When used in a Standard indicates a mandatory feature.

Should: When used in a Standard indicates a recommendation, but is not a mandatory feature.

Slot Velocity: Speed of air moving through fume hood baffle openings.

Smart Hood: A smart hood is one that incorporates a digital controller to manage various functions and performance parameters of the fume hood.

Smart Laboratories: A smart laboratory integrates a number of smart devices that can be controlled with a digital interface.

Smoke Candle: Smoke-producing device used to allow visual observation of airflow.

Smoke Machine: A machine that creates a non-toxic theatrical fog. Used for airflow visualization.

Spot Collector: A small, localized ventilation hood usually connected by a flexible duct to an exhaust fan. This device is not a laboratory fume hood.

Stack: The tube/device on the end of a ventilation system, which disperses exhaust contaminants for dilution by the atmosphere.

Standard Air: Standard Conditions STP Dry air at 70 degrees F, 29.92 in Hg.

Standard Operating Procedure, SOP: This is a formal and written procedure for conducting activities.

Static Pressure: Air pressure in laboratory fume hood or duct, usually expressed in inches of water.

Static Pressure, SP: The pressure developed in a duct by a fan; SP exerts influence in all directions; the force in inches of water measured perpendicular to flow at the wall of the duct; the difference in pressure between atmospheric pressure and the absolute pressure inside a duct, cleaner, or other equipment.

Static Pressure Loss: Measurement of resistance created when air moves through a duct or hood, usually expressed in inches of water.

Suction Pressure: See Static Pressure.

Sulfur Hexafluoride (SF6): Tracer gas widely used for ASHRAE and EN testing. It can be detected with a leak detector to measure loss of containment.

Superstructure: The portion of a laboratory fume hood that is supported by the work surface.

Supplemental Air: Supply or auxiliary air delivered to a laboratory fume hood to reduce room air consumption.

Supply Diffuser: An outlet in the room for the supply air. There are many types, but around fume hoods, a low velocity type is preferred.

Synchronized Supply Fume Hood: This is a new class of fume hood that brings supply air directly into the fume chamber and both the supply and the exhaust are controlled by independent VAV valves so that the proper balance with the fume chamber can be maintained. As the exhaust volume goes up and down, the supply is synchronized to track it.

Synthetic Sensors: General-purpose sensing, wherein a single, highly capable sensor can indirectly monitor a large context, without direct instrumentation of objects. A super sensor that can indirectly monitor a number of activities and provide actionable output.

Table Top Hood: A small, spot ventilation hood for mounting on table tops. Used primarily in educational laboratories. This device is not a laboratory fume hood.

Temperature Gradient: A temperature gradient is a physical quantity that describes in which direction and at what rate the temperature changes the most rapidly around a particular location.

Fume Hood Test Room or Test Lab: This is a space designed and equipped based on criteria in the testing specifications such as ASHRAE110 and EN 14175. This space is designed to have near perfect room conditions so what is being tested is the hood's performance independent of the room.

Thermal Anemometer: A device for measuring fume hood velocity utilizing the principle of thermal cooling of a heated element as the detection element.

Threshold Limit Valve-Time Weighted Average (TLV-TWA): The time-weighted average concentration for a normal 8-hour workday or 40-hour week, to which nearly all workers may be repeatedly exposed, day after day, without adverse effect.

Titanium Tetrachloride: Chemical that generates white fumes used in testing laboratory fume hoods.

Total Pressure, TP: The pressure exerted in a duct as the Algebraic sum of the static pressure and the velocity pressure.

Total Suspended Particulate Matter: The mass of particles suspended in a unit volume of air (typically one cubic meter) when collected by a high-volume sampler.

Transfer air: Air that moves between spaces in a building, driven by the ventilation system.

Transport Velocity: Minimum speed of air required to support and carry particles in an air stream.

Turbulent Flow: Airflow characterized by transverse velocity components, as well as velocity in the primary direction of flow in a duct; mixing velocities.

Turndown: See setback

TWA, Time Weighted Average: The average exposure at the breathing zone.

Ultra-Low Flow Fume Hoods: This is a class of product that is designed to operate at very low average face velocities. For years the benchmark was 100fmp/0.5mps then the low flow hoods were operating at 60fmp/0.3mps. Ultra-low flow goes below that. Often times supplemental air is injected into the fume chamber in an effort to increase containment.

Variable Air Volume - two-position ventilation system: A constant air volume ventilation system (sometimes also referred to as a "two-position variable air volume system") that is designed to provide two separate levels of airflow. The higher level of airflow is provided when a facility is normally occupied such as during regular work hours. The lower level of airflow is utilized during unoccupied times (e.g., nighttime, holidays, etc.) when ventilation needs and internal loads require less airflow.

Variable Air Volume (VAV): In a HVAC system, the supply air volume is varied by dampers or fan speed controls to maintain the temperature; in hoods, the exhaust air is varied to reduce the amount of air exhausted. If the fume hood is going to operate in VAV mode, then the room supply has to also be VAV to keep the room in balance.

Variable air volume (VAV) ventilation system: A type of HVAC system specifically designed to vary the amount of conditioned air supplied and exhausted from the spaces served. The amount of air supplied and intended to meet (but not exceed) the actual need of a space at any point in time. In general, the amount of air that is needed by a space is determined by the required rate and the amount of airflow necessary to maintain comfortable conditions (temperature and humidity). In laboratories, we also consider the required number of ACH to minimize the risk of exposure to the users.

Variable Frequency Drive (VFD): Is a type of adjustable-speed drive used in electro-mechanical drive systems to control AC motor speed by varying motor input frequency. These are very common on fans used for exhaust and supply air.

Variable Volume Hood: A hood designed so the exhaust volume is varied in proportion to the opening of the hood face by changing the speed of the exhaust blower or by operating a damper or control valve in the exhaust duct.

Velocity, V: Magnitude of air motion perpendicular to the plane of the airflow cross section.

Velocity Pressure: Pressure caused by moving air in a laboratory fume hood or duct, usually expressed in inches of water.

Vena Contracta: is the point in a fluid stream where the diameter of the stream is the least, and fluid velocity is at its maximum, such as in the case of a stream issuing out of a nozzle (orifice). It is a place where the cross-section area is minimum.

Ventilated Enclosure: These are often fume hood like devices that are used for exhausting laboratory air but do not meet the definition or performance criteria of a laboratory fume hood.

Visualization of Airflow: Since airflow can't be viewed directly, we have developed various methods to help visualize how the airflow is behaving. The most common is theatrical smoke or fog, also smoke bombs, smoke candles can be used. Using a modified laser beam can further enhance the visualization by projecting it onto the smoke. This is helpful when trying to understand air patterns within the fume hood.

Volume Flow Rate, Q: The quantity of air flowing in cubic feet per minute, cfm, scfm, acfm.

Volumetric airflow rate: The rate of airflow expressed in terms of volume (cubic feet or liters) per unit of time. These are commonly expressed as cubic feet per minute (cfm) in USCS units or liters per second (l/s) in SI units. (Also see room ventilation.)

Vortex: A mass of whirling air, a rotating air mass that often forms insides fume hoods. The most common is a horizontal vortex that forms at the top of the fume chamber behind the sash. But in combination sashed hoods, it is also common to see vertical vortex that form behind the closed section of sash. The performance issue with vortex is that they can collect contaminants and spin them near the sash opening increasing the chance for loss of containment.

Walk-in hood: See floor-mounted hood.

Weather Cap: Device used at the top of an exhaust stack to prevent rain from entering the stack end. Because of the need for a high velocity discharge, only specially designed rain guards should be used. Rain caps that cover the stack are not acceptable.

Work Space: The part of the fume hood interior where apparatus is set up and fumes are generated. It is normally confined to a space extending from six inches (15.2 cm) behind the plane of the sash(es) to the face of the baffle, and extending from the work surface to a plane parallel with the top edge of the access opening.

Work Surface: The surface that a laboratory fume hood is located on and supported by a base cabinet. In the fume chamber, the surface is recessed to contain spills.

Made in the USA
Monee, IL
31 October 2020